Degrees of Freedom

Degrees of Freedom

Living In Dynamic Boundaries

Alan D M Rayner

School of Biology and Biochemistry,
University of Bath

ICP

Imperial College Press

Published by

Imperial College Press
516 Sherfield Building
Imperial College
London SW7 2AZ

Distributed by

World Scientific Publishing Co. Pte. Ltd.
P O Box 128, Farrer Road, Singapore 912805
USA office: Suite 1B, 1060 Main Street, River Edge, NJ 07661
UK office: 57 Shelton Street, Covent Garden, London WC2H 9HE

British Library Cataloguing-in-Publication Data
A catalogue record for this book is available from the British Library.

DEGREES OF FREEDOM
Living in Dynamic Boundaries

ISBN 1-86094-037-4

Printed in Singapore.

PREFACE

Imagine yourself to be completely alone. For most of us the thought is terribly disturbing. We know that if we are to survive at all, let alone lead fulfilled lives, we have to cooperate with others—in couples, families, teams and communities. On the other hand, to cooperate inevitably reduces our freedom as individuals to choose to do what we alone might want, and to rise "above the common crowd". For many people such individual freedom is one of the most cherished of human notions and something well worth fighting to preserve.

An apparent conflict of interests therefore arises between what we may desire as individuals and the needs of others in the groups to which we each belong—whether within our homes, workplace or neighbourhood. In asserting our own freedom to do as we please, we frequently deprive others, for example, by taking all the best bits from a buffet lunch, playing music too loud or driving recklessly. By contrast, if we yield to others' demands we may diminish our own aspirations and risk losing our individual "rights".

This conflict has laid siege to our consciences throughout human history, accompanying the rise and fall of civilizations and causing instability in relationships of all kinds. We are all familiar with the disruption and distress that can be caused by those people who put their own personal advancement above the welfare of others. At the same time we may despair at the "weak will" of those who allow themselves to be steamrollered into subservience, whilst admiring the spirit of others who stand up to an oppressor.

The anguish that this self-or-group conflict can bring comes partly through thinking that some absolute choice has to be made between collective and individual actions. This idea is central to all kinds of extreme political, religious and even scientific dogmas. The one feature that all these dogmas have in common is the derisive way in which their adherents treat any sort of in-between solution as compromise.

Such thinking may have been reinforced by the tendency for human beings to have lost sight of their biological roots, particularly in "western" civilizations. Many people all too often regard human beings as separate from other forms of life on this planet, especially with respect to our ability or "free will" to make conscious choices. They therefore feel that the self-or-group conflict is somehow uniquely human and that there are few lessons to be learned from elsewhere in nature.

The aim of this book is to show that the tension between self and group can be immensely creative and is neither a cause for making absolute choices nor unique to human individuals and societies. Fundamentally the same sorts of unstable relationships that occur in human societies can also be found in what most people would think of as utterly different and unconscious forms of life, such as plant roots, ant swarms and the cellular networks formed by fungi. These relationships can also be found outside the living realm—in river basins, lava flows and chemical reactions, for example. Collectivism and individualism, association and dissociation, reaction and diffusion, constraint and freedom, yin and yang—however these polarities might be referred to in words—are truly primal modes of being. Furthermore, neither of them can exist in isolation; in a deep sense they are complementary and interdependent rather than conflicting. It is the interplay between these polarities rather than the selection of one or other of them that is responsible for the rich diversity of life. The varied forms which emerge from this interplay depend extraordinarily sensitively on local circumstances that dictate where the balance, not compromise, is struck between associative and dissociative processes.

This way of viewing the diversity of life as the outcome of a variable balance between individual and group actions is ancient, not modern, and finds all kinds of expression in literature, mythology and religious belief. So, why should I feel impelled to add another drop in the ocean? Perhaps it is to exorcize my own sense of frustration. It saddens me to witness the continuing confrontational postures of those who regard competition or coercion as the only possible means of social regulation and advancement. I am dismayed by the divisions that have developed between the arts, humanities and sciences, and between reductionistic and holistic approaches to understanding nature. Confrontation and intolerance can only aggravate the energy-sapping stress that besets the lives of so many human individuals and communities. By contrast, sympathetic understanding and mutual support, as opposed to unthinking acceptance or rejection, can alleviate this stress and so allow a more fulfilling and less wasteful existence.

I think that such understanding can be enhanced by becoming more aware of the extent and origins of the diversity of life forms. It is then possible to appreciate just how deeply-rooted and creative the self-group interplay really is. I will therefore try to illustrate the enormous extent of this interplay by drawing examples from across the full spectrum of life forms that inhabit our planet, ranging in size from groups of molecules to social organizations containing millions of individuals.

Some of the examples I will use arise from my personal experience of working with fungi (mushrooms, puffballs, bread moulds etc). The fungi are an entire kingdom of organisms: their total weight may well exceed the total weight of animals by several times and there may be many more species of them than there are of plants! In many natural environments fungi provide the hidden energy-distributing infrastructure—like the communicating pipelines and cables beneath a city—that connects the lives of plants and animals in countless and often surprising ways. Yet the fundamental relevance of fungi to understanding how life forms emerge and change appears to have been strangely neglected. Whilst I deeply regret this neglect, it may at least give me a chance to catch you by surprise, by describing an unfamiliar case that causes you, however fleetingly, to view life in a different way.

What fungi have taught me to appreciate is the enormous significance of indeterminacy or "open-endedness" amongst all kinds of life forms. Indeterminacy is due to the continual interplay between association and dissociation that results from the possession of a dynamic boundary between a system (something containing matter or energy) and its surroundings. In fungi, this interplay generates varied branching patterns in the interconnected, protoplasm-filled tubes (hyphae) that spread through and absorb sources of nutrients. The hyphae branch away from one another (i.e. dissociate) most prolifically when nutrients are freely available, but reassociate to form such structures as mushrooms when supplies are depleted.

Indeterminate systems are able both to change and be changed by their local environment and cannot be understood as simple assemblies of fully discrete, building-block-like units. They are therefore continually creative as they gain, distribute and discharge energy, and their long-term future cannot be predicted—no matter how much is known about them at a particular instant in time.

Even so, an enormous amount of thinking about living systems continues to be based on the assumption that they are made up of discrete, predictably interacting units. It is also supposed that these units, which are commonly described as genes, cells and individuals, can simply be discarded by natural selection if they don't fit in. From this perspective, competition appears to be of paramount importance in the refinement of super-efficient, centrally administered organizations. Most fundamentally, genes become cast as central controllers of the temporary contraptions—bodies—that convey them through successive generations.

In that it doesn't take in the widespread occurrence and fundamental importance of indeterminacy, I find this gene-centred, building-block perspective too restrictive. Worse, I suspect that it may be reflected in the mismanagement of human societies by hierarchical, competition-based, central power structures. It is a huge mistake to try to impose centralized order on an indeterminate system such as a society. It is also a mistake that is immensely costly in the pain, demoralization, distrust and squandering of resources that it causes.

My reasons for saying all this lie ahead. I should say straight away that I don't want the book to be a backward-looking review of the work of those many people who may have had similar thoughts or made relevant observations. I shall therefore refrain from citing innumerable literature sources, although I will provide a list of texts for further reading at the end—without claiming to have read, understood and agreed with them from cover to cover! The perceptive reader will detect vast gaps in my knowlege and probably in my understanding, as well as a tendency to re-invent the wheel. Specialists will find their areas of interest referred to in ways that may seem indirect or over-simplified. I hope, though, that they may be stimulated by seeing their knowledge applied in an unfamiliar light.

Alan Rayner

Bath

September, 1996

A READER'S PREDIGEST

I want this book to appeal to anyone who is interested in *thinking* about and trying to *understand* the varied ways in which life forms organize themselves.

For those who do not have a strong scientific background, I want to dispel some of the rigidity and mystique that can make science seem so cold and inaccessible. I want to show how scientific perspectives can create spiritual as well as material wealth.

For those analytical scientists who prefer to reduce life into discrete "building blocks", I want to explore the complementary benefits of a more fluid approach that views life within the context of dynamic boundaries.

For those who don't have a strong biological background, I would like to demonstrate the enrichment of appreciation and understanding, not least of ourselves, that comes with an awareness of the diversity and commonality of living things. Equally, I want to explore the concept that the patterns formed by living things are not unique in themselves, but involve universal dynamic processes of counteraction between association and dissociation.

For those biologists who focus only on particular kinds of organisms or particular aspects of living processes, I want to demonstrate the benefits of taking a more encompassing view that includes unfashionable or neglected subjects.

To try to do all this, I have to provide the bread of information as well as the butter of ideas. Ideas alone are slippery and their relevance can only really be appreciated when they are related to specific knowledge of substance. Substance alone is indigestible in its mass of detail; in trying to come to terms with this detail it is impossible, without some conceptual awareness, to distinguish underlying general truths from secondary ornamentation—i.e. to see the wood for the trees.

The ideas expressed in this book originate from a personal quest to identify the dynamic pattern-generating processes that underlie the diversity of living forms and behaviours. In developing these ideas, I have surveyed as much biological diversity as is "visible" to me and, whilst trying to avoid prejudice, looked for recurrent features. These recurrences, when (and if) found, have then enabled me to identify basic themes.

In the first chapter, I aim to show how important the way we identify and define boundaries—beginnings and endings—is to thinking about how nature is organized. If, as many people suppose, living systems truly are wondrously sophisticated, machine-like assemblies of discrete components, it is necessary for boundaries to be defined with extreme precision. I will argue that such precision is unattainable because living system boundaries are dynamic. I will also suggest that to imagine that such precision exists can result in misunderstandings and paradoxes, even though in the short term it can certainly help us to make limited calculations and predictions.

The second chapter provides some basic biological information. It consists of a whistle-stop tour of the diversity of life forms, pausing to reflect on the importance of association-dissociation interplay at all levels of organization.

Having hopefully by then set the scene by outlining the ideas and information on which the book is based, the subsequent chapters will explore individual issues in more depth. In the last chapter, I will try to draw some conclusions about how a biological perspective may help in understanding and perhaps even improving human relationships.

I have provided as much detail as I hope is necessary to allow the grounds for my interpretations to be understood and evaluated. Where I think that the interpretations are subtle or complex, I have tried to avoid taking easy "short cuts" that could prove to be misleading in the longer term. I must therefore ask for the reader's patience in not expecting everything to be clear without further thought—in fact if my writing provokes thought, I will be delighted. I have also introduced some technical terms as and when appropriate. I know that "jargon" can make a topic inaccessible, and I suspect that some people use it deliberately to demarcate intellectual territory—to warn off would-be competitors. However, it is sometimes very helpful to coin a special word or phrase to describe something particular, and some readers may also like to be made aware when such terms exist. I have therefore tried to use technical terms as unobtrusively as possible—in ways that allow them to be glossed over if desired—and only when I think that there is a definite value in doing so.

ACKNOWLEDGEMENTS

There are many people without whose encouragement, technical help and constructive criticism I could not have proceeded. First there are all those people who have assisted me directly during my researches on fungi, by doing experiments, making observations, arguing, delving into the literature and providing the innumerable technical skills that I have never managed to acquire personally. Most recently, these people have included Martyn Ainsworth, John Broxholme, David Coates, Jon Crowe, Fordyce Davidson, Chris Dowson, Gwyn Griffith, Roger Guevara, Jo Kirby, John Mukiu, Mark Ramsdale, Richard Scrase, Priscilla Sparkes (née Sharland), Christian Taylor and last but by no means least, Zac Watkins. Then there are those, both inside and outside fungal biology, who have given scientific support from further afield. Here I would especially like to mention John Beeching, Lynne Boddy, Don Braben, Clive Brasier, Tom Bruns, Nacho Chapela, John Crawford, Juliet Frankland, Nigel Franks, Monique Gardes, Tom Gordon, Steve Hendry, Mike Mogie, Martin Pearce, David Ray, Karl Ritz, Ana Sendova-Franks, Ed Setliff, Brian Sleeman, Jan Stenlid, Norman Todd, Jitendra Vaidya and last, but by no means least, John Webster. Finally there is my immediate family: my wife, Marion; my daughters, Hazel and Philippa; my parents, Ronald and Mervyn; my sister, Joy, and my parents-in-law, Beryl and Ted who have witnessed first-hand the trials and tribulations of trying to get my thoughts into an intelligible form. I thank them all.

CONTENTS

CHAPTER 1

DEFINING DYNAMIC BOUNDARIES

1. Uncertainty

The need to define boundaries—to know precisely where one thing ends or another begins—is fundamental to the way many of us think and act. An awareness of boundaries prevents us from stepping off the edges of cliffs and allows us to separate things from one another and place them in different containers or categories. It also enables us to focus our attention and efforts on all kinds of arenas, from fields of view down a microscope to sports stadiums. By contrast, a world consisting only of fuzzy boundaries, where everything grades into everything else, would seem intolerably vague and uncertain. We would be left to grope around not knowing where we are or what we are doing.

It comes as no surprise, then, that we attempt to assure ourselves of a secure and predictable passage through life by spending a huge amount of our time dividing up and classifying our surroundings. We think that learning how to recognise boundaries and to appreciate their significance is one of the most important aspects of growing up.

In order for our children to mature into reasonably balanced, adult members of society, we therefore try early on to provide them with firm guidance about what is around them, their place in the scheme of things and what is or is not acceptable behaviour. At the same time, a universal characteristic of children is that they continually probe and test the limits of their surroundings—not least the tolerance of those curiously unpredictable beings, their parents! When the limits are found, particular kinds of behaviour and interpretation can often be reinforced into persistent attitudes. We tend to become domineering or subservient, callous or sensitive, sceptical or gullible, cynical or admiring, conservative or radical—depending on the interaction between our experiences and individual genetic make up. As this reinforcement occurs, so our ability to change attitude in response to new experiences progressively declines and we may become fixed in our ways. Even so, we tend to hold on to our childishness, and the openness to new possibilities that comes with it, for a far greater proportion of our lives than other mammals, including our nearest ape relatives. This adventurous tendency is fundamental to our pioneering spirit as a species.

Scientific knowledge, meanwhile, is popularly thought to advance securely and predictably through the gathering of numerical "data" (units of information) by counting and making measurements. To achieve this it is necessary to be certain where the boundaries of the units being counted or measured are. Sometimes this seems easy, as when assessing the size of a resting flock of birds. At other times it gets difficult, as when the flock takes off and keeps splitting up, recombining and expanding and contracting its boundaries. In spite of such difficulties, calculations based on precise data-gathering can sometimes allow extraordinarily accurate predictions to be made. For example, they can allow us to plot the progress of a chemical reaction or the movements of celestial bodies. The ability to make such accurate predictions brings us deep satisfaction and reinforces our faith in the methods of exact science.

Encouraged by the successes of data-gathering and analysis, many scientists have been won over by the power of disciplined, precise quantification as the means of understanding nature and turning it to technological advantage. They view the assembly of building blocks of concrete facts as being of great importance in the construction of a sound knowledge base and find human uncertainty and error-proneness difficult to tolerate. Recently, there has even been a trend to liken the workings of the human brain and the pattern-generating processes of life itself to the operations of a *digital* computer on *discrete* "bytes" of information, encoded in genes.

On the other hand, amongst the general public, there are signs of a growing disenchantment with science. Its seemingly hard-edged, unadventurous, intolerant face does not appeal to the sensitive, excitable elements in human nature—least of all to the young who are its supposed heirs. Whilst science, when applied as technology, may do much to improve the material welfare of human beings, it is rarely perceived from outside as doing much for the soul. It may warm our living rooms but not our hearts! Also, by justifying their work purely in terms of practical benefits and "quick fixes", many scientists have made rods for their own backs. They now find themselves at the beck and call of unrealistic public expectations, prevented from doing curiosity-driven research and being blamed for providing the weaponry of war. Woe betide those scientists who do not come up with clear predictions, provide immediate solutions to practical problems, or foresee the negative implications of their discoveries. A loss of face in scientists is a loss of faith in science. What better recipe for secrecy, protectionism, censorship, authoritarianism and the division of societies into scientifically well-informed and scientifically alienated subsets?

Science should not, and need not be perceived like that. Based most fundamentally on observing, recording and interpreting phenomena, with the *aid* of experiments and measurement, scientific understanding is rewarding spiritually as well as materially. It is not in conflict with, nor does it replace the need for the arts and humanities, though some of its current conclusions may contradict certain aspects of religious doctrine. It is capable of by-passing and even benefitting from the kind of human error which is in fact a vital component of having imagination and being able to adapt to unpredictable circumstances. *It is as much about uncertainty as certainty.* To acknowledge uncertainty and to recognize its origins, is scientific strength, not weakness; to insist on certainty is arrogance, not strength.

Uncertainty brings the spice of adventure into scientific endeavour. It partly arises, and will always do so, from incomplete knowledge. Even more fundamentally, it is due to the impossibility of defining boundaries absolutely. This impossibility becomes especially apparent when the ability of a system (something containing matter or energy) to be *dynamic*—to change with the passage of time—is taken into account. The absolute absence of boundaries means, literally, that nothing can be perceived to exist. On the other hand, an absolute boundary—one that is completely sealed—permits no exchange of energy between a system and its surroundings. At least in the world of everyday experience, any such fully "isolated" or "self-contained" system would be unable to gain or lose anything and so its outward form could not change with time. In fact, no isolated systems are known, although it is commonly assumed that the universe itself must be one. Even the solid rock of mountains is generated by active geological processes and eventually crumbles, though the time scale over which the underlying exchanges of energy occur are immensely protracted by human standards.

If boundaries cannot be defined absolutely, at least in the long term, then the notion that fully discrete units exist at all cannot hold true, and everything must be to some degree in a state of flux—the more so the less solidly it is bounded. Discreteness is therefore an abstraction, an unattainable "ideal". However, our ability to assess quantities and make calculations implies discreteness in that it depends on being able to define the boundaries of the units being counted or measured. In the real world, and perhaps the real universe, the most that our calculations may therefore be able to achieve is a very close approximation to or simulation of the truth, and then only by the device of inventing infinitely large and infinitesimally small numbers. When this point gets forgotten, there is a danger of treating numbers and quantities as absolute realities, even though boundaries cannot be totally hard and fast in any dynamic systems. This can badly distort and limit understanding, especially of living systems.

Correspondingly, being alive—thriving as well as surviving—depends most fundamentally on taking in the energy supplied in sunlight or food, distributing it to good effect, and withholding it from the outside world, and to do all this life forms must have *dynamic boundaries*. These boundaries form *reactive interfaces* that can open, close, expand, contract, relocate and take one another in as energy supplies wax and wane—from the membrane of an amoeba that engulfs a bacterium to the great maw of a whale that takes in a shoal of fish. Think of a gannet plunging into the ocean to seize energy in the form of a fish, then returning to its usual home in the ocean of the air. Think also of a plant unfurling and shedding leaves, and then of a caterpillar munching its way through the foliage, growing and moulting—and then pupating and finally re-emerging as a butterfly, allowing it to transfer from the pedestrian world of the leaf into the world of air and flowers. You should be beginning to get the picture.

Discrete as an individual animal such as an amoeba, an insect, a bird or a mammal might appear to be at an instant of time and irrespective of its neighbours, its living envelope—its boundary—creates and follows innumerable interconnected highways and byways as it moves through space and time. Also, the same kinds of patterns that result from an animal's movements are formed by the body boundaries of many plants and fungi that are capable of growing and changing indefinitely. The properties of these living boundaries are due to the interaction of ingredients coming both from inside and outside the system—not least the water that envelopes the genes of all living forms.

It is therefore at boundaries that all life's action occurs—the places where nature (genetic influences) and nurture (outside influences) combine and inextricably intertwine to generate the rich complexity of the living world. These boundaries can never be completely fixed, but instead define the ever-changing *contexts*, the local environments within and between which life processes are transacted across scales of organization ranging from microscopic to global. When it comes down to it, life forms are moulded within dynamic contexts containing *wet genes*, and not assembled from dry building blocks.

Nonetheless, many people try to understand and predict the behaviour of living systems using precise definitions of the boundaries of individual organisms and their components. These precise definitions are critical to the way that many fundamental concepts have been developed and applied. These concepts include *competition*, the demand of two or more individuals for the same resource; *coexistence*, the simultaneous presence of two or more individuals; *multicellularity*, the co-existence of cells in the same individual; *symbiosis*, the persistent, intimate association of individuals, and *natural selection*, the genetic survival of one individual rather than another.

Precise definition of boundaries is also the basis for rigidly hierarchical schemes for classifying organisms or groups of organisms either by name (i.e. "taxonomically") or in terms of their distribution in natural environments (i.e. "ecologically"). Correspondingly, individuals are identified as the component units of "populations" and populations are the component units of "species". Taxonomically, species are the components of "genera", genera the components of "families", families the components of "orders" and so on. Ecologically, species and all larger taxonomic groupings are components of "communities", and communities are components of "ecosystems" which are components of the "biosphere"—the entire realm of living things on the planet.

However, since there are no absolute criteria for making such definitions, much heat has been lost in pointless debate about where to draw the lines! It is fine to have *signposts* telling approximately where you are in the general spread of variation, like the red, orange, yellow, green, blue, indigo and violet colours of a rainbow. However, to expect these signposts to give absolutely precise directions is to court humiliation. Since a loss of face leads to a loss of faith, the authors of precise definitions have a history of being inordinately self-protective, arguing in legal rather than scientific terms and impeding understanding.

2. Individuals and Collectives

To return to what many people regard as the basic "unit" of biological diversity—what *exactly* is an individual? It should now be clear that this deceptively simple question has no simple answer because throughout the living realm, as elsewhere, the discreteness of individual entities is blurred by their tendency to associate with others in dynamic boundaries. What might seem from one point of view to be an entity in its own right can always be seen to be composed of smaller sub-units or as a component of a larger system, when looked at in another way. Societies are assemblages of individuals; individuals can be assemblages of organs; organs are assemblages of cells; cells are assemblages of organelles; organelles are assemblages of molecules—and so on.

Rather than asking *what* an individual is, it therefore makes more sense to ask *how* individual some entity is. To decide on this depends on the degree to which the entity is connected to or disconnected from others, and to what extent it can be considered to be a "part" of a larger something or a "parcel" of smaller somethings. These issues depend not only on where boundaries between entities are located but also on the "frame of reference" or *context* within which these entities are being observed.

For example, think about a wide-necked bottle containing two glass marbles. Viewed from fifty metres, only one boundary will be apparent, that of the bottle. At this range, the bottle is the individual. Now view the bottle from inside; only the marbles appear as entities. If the bottle is shaken erratically, so energizing the system and overcoming the attraction of gravity, the marbles appear to move independently for most of the time. Yet they never get more than a certain distance apart, as if they were joined by an invisible string.

On examining the bottle from about a metre, this curious behaviour is all too readily explained. The marbles co-exist within the context of the bottle boundary but are discrete with respect to one another. They are separate and yet not separate, free from each other and yet constrained by the boundary that surrounds them both. Furthermore, if more marbles are added to the bottle, they will increasingly constrain one another and start to behave dynamically as if they were a coherent entity in their own right; viewed from a distance they may even appear to be a uniform mass. However, when poured out onto a smooth, hard horizontal surface such as a polished table, their individuality is immediately reasserted as they disperse in all directions. By contrast, when poured onto a yielding surface, for example a tray of moist sand, the interplay between their own tendencies to disperse and the resistance of the surface generates a diversity of patterns. These patterns emerge because the marbles both impede one another and create paths of least resistance for others to follow. Exactly the same kinds of patterns would be produced if, for example, a crowd of people were to try to race across a field of tall grass.

Now imagine half the marbles to be black and half of them white and that they are arranged in two layers within the bottle. There is now yet another boundary to consider; at the interface between the layers. This boundary is visible even at long range. If the bottle is shaken, then as long as there is at least some freedom of movement, the layers will begin to merge. Eventually it will seem that the marbles are fully mixed, that there is no longer any boundary separating black and white, and that indeed from a distance the marbles form a uniform grey mass.

No matter how long the bottle is now shaken, the situation will no longer seem to change; the boundary between black and white will never be restored—unless by some extremely improbable fluke. However, closer examination reveals that the boundary between black and white has not ceased to exist, because the marbles have not actually coalesced. Rather the boundary has become very much more complex. It can therefore only be viewed at those small observational scales where the larger picture of overall relationships within the bottle boundary are not evident.

The preceding discussion should hopefully have shown how important scale of observation and context are, and how easily paradoxes can arise if the inherent difficulties in defining boundaries are not appreciated. These paradoxes have caused enormous confusion in environmental and evolutionary biology, as well as in other scientific and social fields.

The discussion should also have shown how even such hard, inanimate, incohesive entities as marbles can exhibit life-like interrelationships if they are mobilized in some way at the same time as being subjected to some external constraint. These interrelationships can only result in shifting patterns if the constraint is not absolute. The bottle must not be crammed full. It must be capable of being shaken, allowing external energy to be transmitted within its interior. At least at some time in its history it must have been open for the marbles to get inside and must be opened again for them to get out. When the marbles are poured onto moist sand it is the "give" in the sand and the resultant *deformability* of the enlarging boundary between marbles and sand, that is critical. If a boundary is deformable, then the form it takes will depend both on its own inherent properties and on the behaviour of the entities it constrains. It will both define and be defined by the system. A system cannot be defined as existing independently of its surroundings or *vice versa*. Definitions of system and boundary are as interconnected as are chicken and egg.

The imposition of an external constraint is not, however, the only way in which mobile entities associate into patterns and lose their individual integrity. They may also be constrained by their attraction for one another—in fact, such attraction may suggest the existence of some "invisible" boundary, some interdependence or "field" interconnecting their activities. Imagine that magnetised steel spheres had been used instead of glass marbles to occupy the bottle boundary discussed earlier. The dynamic relationships between the spheres now depend on the extent to which their mutual attraction is counteracted by processes, such as shaking or opening the bottle, that tend to cause them to separate. If the attraction is too strong, the system will become resistant to change; no spheres will escape from the bottle when it is upturned unless some disruptive process is initiated.

Though the actual mechanisms and their scale of operation may vary greatly, such counteraction between attraction and repulsion of individual entities causes patterns to form in all kinds of living systems, including human societies. However, to understand this point fully, there is one more stage in the argument that needs to be developed.

Like the marbles, the steel spheres (as opposed to their magnetic fields) remain physically discrete even when they are closely associated. However, in many systems, both living and non-living, the end result of physical association is *integration*. This can occur through the *enclosure* of a smaller boundary within a larger boundary, so creating "part and parcel". Alternatively, it may involve *coalescence*, such that the contents of formerly discrete boundaries become continuous, so converting formerly separate parcels into a single parcel.

Integrational processes allow not only the persistence and repetition of particular patterns as "variations on a theme", but also introduce entirely new themes on new scales. These new themes themselves can then be varied by the counteraction of association and dissociation. Feeding on its own creativity, an evolutionary process is set in train that generates diversity on potentially countless organizational scales.

3. Self, Nearself and Nonself

What sorts of entities can integrate, and what are the consequences of integration for their individual survival? These questions connect the problem of deciding what is an individual to the problem of deciding what is a self.

It has been argued, forcibly, that self-sacrifice purely in the interest of others—"non-reciprocal altruism"—is quite literally an evolutionary dead-end. In a competitive world where not all can survive, "group selection", in which the interests of groups are served at the expense of individuals, can exist only transiently, if at all.

At first sight, this conclusion might seem to suggest that collectivism (living together) is not a viable route to evolutionary success. However, it is wrong to equate individualism (living apart) with selfishness and collectivism with altruism. Ultimately, all persistent kinds of associations between living entities originate for one or more of three non-altruistic reasons. First, harking back to the marbles in a bottle, the entities may in effect be "prisoners", confined within some external boundary that they cannot escape from—like bacteria in a test tube or fish in a pond.

Second, they may have common needs or "self-interests" and so tend to congregate in the same locations—like the audience at a concert or spectators at a football match. Third, they may be able to gain some form of protection or succour from their associates—like street gangs or house guests.

Collectivism may therefore have its roots just as much immersed in "self-interest" as does individualism. However, as will be discussed further in Chapter 7, it can give rise to non-self-interest. It can also, sometimes literally, bear fruit of a radically different kind from individualism—as when a mushroom springs up from a fungal network.

But what is "self-interest" if the individual cannot be defined with absolute precision? What is self, what is not self (nonself) and where are the boundaries between them? These questions introduce the need to understand the reciprocal relationship between competition and communication, and to recall the fundamental nature of life forms as energy-using systems.

Where two life forms co-exist, that is they are constrained by some physical or functional boundary external to them both, they are liable to compete for resources. If there is inadequate resource to sustain them both, one or even both must cease to function, dissipating the energy they once held within their boundaries. Competition for inadequate resource supplies is not good for mutual survival!

Competition is most intense when the participants have identical demands for resources, i.e. when they are most like one another. If the participants come into actual physical contact, they may even compete for resources *within* one another's incompletely sealed boundaries. If there is the tiniest initial disparity between them, the most active competitor will then draw resources from the less active one. In such ways tumour cells may parasitize their neighbours and big seagulls may steal pieces of bread from the beaks of smaller seagulls.

On the other hand, if the boundaries between the participants can be made to coalesce in some way, communication will replace competition and resources will be shared around the unified system according to demand. Also, the enlarged boundary results in a reduced surface area in relation to volume and therefore reduces losses to the surroundings, as is exploited by a huddle of penguins keeping warm. Collective organization may also enable a protective front to be maintained against would-be consumers, as in a party of wildebeest, and at the same time allow co-operative resource capture, as in a pride of lions. Furthermore, prospects can arise for different entities within the combine to specialize in different activities and hence divide labour in a super-efficient system—as between the heart, lungs, guts and nerves of a mammal or the reproductive, soldier and worker "castes" in a social insect colony.

Clearly, considerable benefits can be gained by like entities—whether they are cells, multicellular organisms or indeed colonies of multicellular organisms—if they combine forces, stop competing and change their scale of operation.

However, there are also dangers in living together. There may be insufficient resources to be shared amongst all members of the collective. The ability to capitalize on plentiful resource supplies may be limited by the reduced surface area in relation to volume, so that the collective is satiated too readily when there is surfeit, only to starve when there is shortage. An enlarged boundary may present a more obvious target to a predator. There may be difficulties in eliminating toxic wastes that could easily be released into the environment in more dispersed systems. Disease and dysfunction can spread more readily amongst members of an integrated system. Should one component in a highly interdependent system fail, the whole system will be brought down.

Where like entities are involved, there are therefore good reasons both for living together and for living apart. This point is beautifully illustrated by two groups of organisms to be described in more detail in Chapter 2, the "cellular slime moulds" and the "true fungi". These organisms can develop both as separate entities when external resource supplies are plentiful and as functionally integrated assemblies when resources become limiting.

The concept of self, which might normally be thought to apply just to any separate individual, may therefore need to be extended to include all entities capable of integration because of their likeness to one another. For living systems to be alike, it is necessary for them to have the same *information* content, in the form of *genes*. Having the same genes is therefore tantamount to belonging to the same self, *providing* that communication channels are kept open or at least openable. On the other hand, if communication channels are absent or closed, then even (or, rather, especially) entities with identical genes will compete, a fact which may lie at the basis of many kinds of cancers and degenerative processes.

At the other extreme are "nonself" entities so very different from one another in their properties and genetic make-up that they do not compete as long as their boundaries remain discrete. If the boundaries of such entities become integrated, generally by a process of enclosure, then some very important possibilities arise. Where the boundary of the smaller entity is not maintained, then this smaller entity is *consumed* by the larger one. If the boundary of the smaller entity remains intact, however, then this entity can draw resources from, exchange resources with and/or be sheltered within the larger one which becomes its "host".

As will be described later in this chapter, these possible kinds of persistent association between nonself entities give rise to "parasitic", "mutualistic" and "neutralistic" "symbioses". In those cases where the host may be damaged by the association, its chances of survival will increase if it can reject the invader in some way. By contrast, where the properties of the combined entities complement one another, then they form partnerships able to thrive in fundamentally different ways from each partner in isolation. Examples of these partnerships or mutualistic symbioses will be described in Chapter 2. They include those formed between fungi and algae in lichens and between bacteria and plant cells in the "root nodules" of the pea family of plants. A particularly striking example is found in the guts of certain termites. Here a symbiotic protist (single-celled organism) aids digestion for the termite by breaking down plant cell walls, and is itself propelled by the action of whip-like "flagella" of symbiotic bacteria embedded in its boundary. The termite is a veritable parcel of parcels!

At first, the original identities of the partners in mutualistic symbioses may be clear from their ability to dissociate from one another and even to sustain an autonomous or "free-living" existence. However, with time, the partners can become increasingly interdependent; in this way entities that were once extreme nonself become part and parcel of the same self. One of the most important events in evolution that is thought to have come about in this way was the origin of so-called "eukaryotic" cells—the kind of cells found in animals, plants and fungi.

Between the extremes of self and nonself are situations where integration occurs between entities that differ in at least some respects, but also have a lot in common. Here the potential mutual benefits of an alliance are balanced against the risks and prospects of one-sided gains and losses or the possibility of mutual interference and degeneration.

This situation can be likened to that between two industrial companies with similar, but not identical products (Figure 1.1). From the point of view of the management of each company, the other company is a competitor, using up resources and having a market share or territory that would otherwise be available to enhance growth and profit margins. If communication channels are opened with the other company, access can be gained to its resources and territory. Through reduction of wastage and mutual exchange of expertise or equipment, it may even be possible to increase overall efficiency and initiate new ventures. On the other hand, there are dangers of takeover by the other company, the inheritance of any industrial relations problems it suffers from, and of incompatibility of machinery or operational procedures.

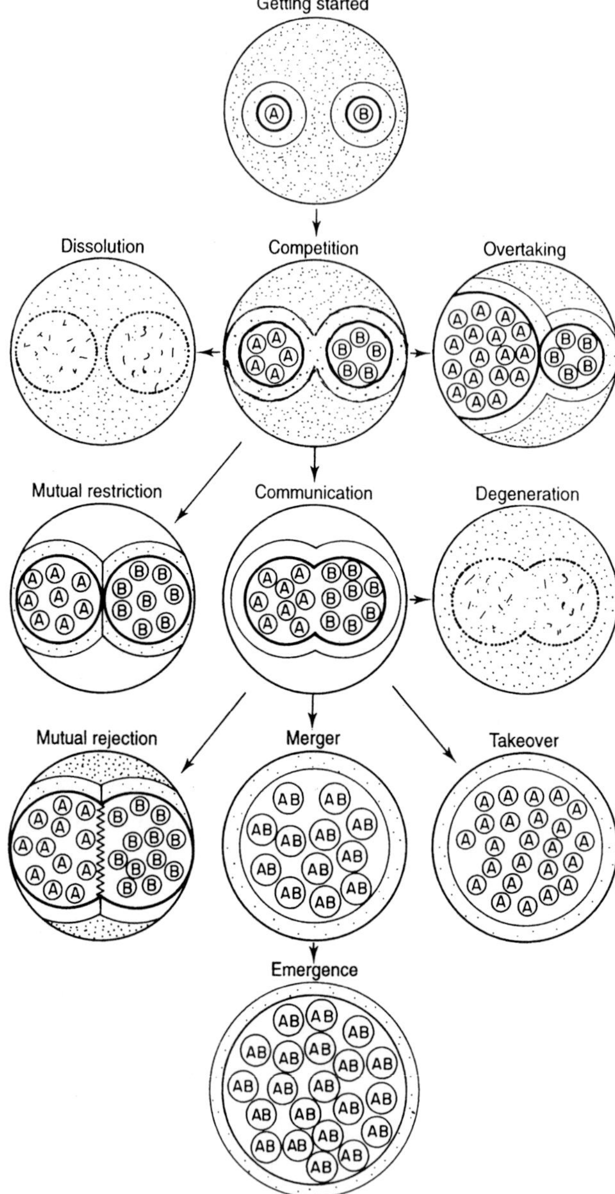

Figure 1.1. Corporate interplays. Two organizations, containing inhabitants (A and B) with similar but not identical expertise and requirements for resources begin to establish themselves in the same arena, represented

here as a circle with densely stippled contents. Each organization is surrounded by a region of "influence" (lightly stippled) which it draws on to support its activities. If the net income from this region balances net expenditure, the organization will maintain but fail to expand its boundaries. If there is a "profit margin" (income exceeds expenditure), the organization will grow. However, if there is a deficit, the organization will lose viability.
Both organizations grow until their regions of influence overlap and begin also to become limited by the boundaries of the arena. The resulting competition can reduce profit margins to the point where no further expansion is sustainable (leading to mutual restriction), or beyond this point and into deficit (leading to dissolution). However, if one organization proves to be more competitive, it will begin to monopolize the arena as it erodes the other's region of influence (overtaking).
On the other hand, if the organizations open their boundaries to one another and start to communicate, they will have the opportunity to operate as one and so expand their mutual influence. Much now depends on the way the inhabitants interact. If they compete or interfere, the resulting incompatibility may culminate in extensive degeneration, mutual rejection or takeover. If they complement, they may not only allow the corporation to fill the arena to its fullest potential, but also enable it to expand the boundary of the arena itself, by means of innovative interactions.
The same organizational principles apply in all kinds of living systems that possess dynamic boundaries, from cells to societies. (Modified from Rayner, 1996a)

The dangers of nearself associations arise directly from the very disparity which under other circumstances might allow for complementation. It is therefore understandable that initial merger or takeover negotiations often break down and a mutual competitive stand-off is resumed, but this time with the boundaries re-inforced by overt antagonism. On the other hand, if merger or takeover is to succeed, so integrating the boundary of the parent companies, any latent incompatibility has to be overridden, either by agreement or suppression. Having been overridden, this incompatibility may nonetheless retain the potential to be expressed at some later stage in the development of the new organization.

This kind of situation, wherein recognisably different entities are potential collaborators as well as competitors, is widespread throughout the living world, and accounts for many seemingly paradoxical patterns of behaviour. As will be explored in Chapter 7, sexual relationships provide a very vivid illustration of the instabilities which can result from the conflict between acceptance and rejection of prospective or actual partners! Defining where self-interest begins and ends, at first sight a simple matter reducible to distinctions between individual genes, is anything but simple.

To summarize, living entities may be expected to integrate both because of their likeness to one another genetically and because of their non-alikeness. In the former case, there may be limits to the amount of difference, or disparity, which can be tolerated. Beyond these limits, coalescence will be restricted by the expression of incompatibilities. If allowed to spread, these incompatibilities may give rise to extensive degeneration but if localized at boundaries, they preserve differences by preventing the possibility of takeover. Where entities integrate because of their non-alikeness, they are able to complement one another and hence divide labour between

..stinctive, specialized roles. However, if the demarcation between these roles is not clear, the resultant interference may lead to takeover and degeneration. Integration brings the promise of innovation but can also be profoundly destabilizing. The rich diversity of living systems depends in no small way upon this fact.

4. Holism and Reductionism

The impossibility of identifying individuals as absolutely discrete entities should now be apparent. However, it is still possible to define the essential attributes of individualism as functional independence, physical separateness and the preservation of boundaries. Conversely, collectivism implies concerted action and the consequent temporary or persistent integration of boundaries. A creative interplay between individualism and collectivism recurs at all levels of biological, if not physical, organization. All biologists, whether they are concerned with molecules, cells, whole organisms or populations therefore face similar issues when trying to understand the complexities of the systems they study.

However, the universality of collectivistic and individualistic themes has not stirred the imagination of many biologists. Ironically, this is probably due to an imbalance between fundamentally individualistic and collectivistic approaches in science. Science progress is itself an evolutionary process, and its course is profoundly modified by the balance between these approaches.

As has already been implied, the currently predominant approach in science is analytical. It involves the exact measurement and description of a system in terms of its most fundamental components and the rules, or even laws that these components obey. It has generally been assumed that knowledge concerning each component can simply be summed to provide a fully predictive understanding of the system as a whole. The probability that systems can, through interactions, acquire properties very different from the sum of their parts and behave unpredictably has effectively been set aside. Unpredictability, where it occurs, has therefore been assumed to be due to "imperfect" boundaries or insufficiently precise knowledge or techniques. Correspondingly, we have, at least traditionally, continued to train our children using building-block logic which has to be taken and assessed seriously in formal lessons and examinations. Male children, especially, have been given toy building bricks and construction kits to prepare them for the "no-nonsense" world of productive "work". By contrast, the common-sense understanding of context that comes from experimenting with water, sand and plasticine, or getting very different results with the same cooking ingredients has been reserved for playtime and home economics classes.

That there is an important role for such reductionism is undisputable. However, pursued exclusively, it is restrictive rather than inventive and leads to the establishment of fixed boundaries between "disciplines". Those imbued with it are often intolerant of error and mistrustful of novelty, unifying theories and "outsiders". They prefer to think of their own "specialism" as self-sufficient and different. The resulting divisions end in the directionless proliferation of detail, dissipation of energy through mutual competition and stifling of progress through over-specialization.

Holism, by contrast, involves an attempt to understand whole systems in terms of interactive processes within them: interrelationships between components are viewed as being more important than exactly what these components are. Recognition of patterns and parallels, and the putting together of previously unconnected themes is therefore foremost. Holism is therefore disrespectful of boundaries, save the one that defines the context of the particular system being studied. Here, specialism evolves in response to the opportunity for effective division of labour rather than to establish competitive boundaries.

The real or apparent eclipse of holism in modern science has led to increasingly impassioned pleas for its re-instatement alongside or even in place of reductionism. The emergence of Chaos and Complexity theories to explain the predictably unpredictable patterns of change in partially bounded systems represents just one aspect of this resurgence. However, pursued exclusively, holism brings dangers as well as opportunities. Disregard of boundaries brings confusion rather than clarity of thought. Complementation between specialisms evolves into interdependence at the expense of the freedom needed to explore further. Such interdependence produces an inflexible structure that is ultimately susceptible to catastrophic change if part of it, like Achilles heel, proves to be a weak link. Similar patterns may arise by random co-incidence rather than for more fundamental reasons. In a competitive world, the very openness which underpins collaborative scientific ventures invites ridicule and piracy.

The truth is that science progress has not been, and could not have been due to the exclusive pursuit of reductionistic or holistic approaches. Rather it is the interplay between these approaches which has been creative. Units of information are generated piecemeal until such time that they are sufficient, and soundly enough based, to contribute to a new or wider synthesis that brings about a change in understanding, a so-called "paradigm-shift". Such new syntheses are commonly the result of the activities of "non-conformist" scientists prepared or naive enough to err (literally, to wander or depart from) outside the boundaries of their own specialisms and thereby to make discoveries and communicate with others. Here, to err is not only an expression of human individualism; it is vital to progress and requires the support of the community which may itself eventually benefit from it (research funding agencies, take note). New syntheses lead to further rounds of analytical, fragmentative study, laying the basis for re-synthesis and further breakthroughs in a cyclical progression. Such is the dependence of new breakthroughs on pre-existing information that nothing is invented or discovered entirely afresh (Patent lawyers, take note). New knowledge is old knowledge in a different context. Rediscovery strengthens discovery.

In a similar way, creative interplay between association and dissociation of "units" of genetic information in a changeable context underpins biological evolution. In the process, the potential benefits of association for survival and spread of these units, i.e. access to resources and innovative interactions, are set against risks of being consumed, subsumed, destabilized or over-stabilized. Evolutionary processes therefore resemble more the unpredictably changing course of a river than the "yellow brick road" of one discrete step at a time. No-one can know exactly the future path that evolution, any more than science progress, will take, especially in the longer term. However, anyone can predict that the organization of life forms will continue to change due to the interplay of individualism and collectivism on a shifting environmental stage. It is also possible to have some inkling of the kinds of *feedbacks*—self-enhancing and self-containing processes—that direct this change.

5. Stage Directions—Opportunity and Competition

For the past century, "natural selection" has been the mechanism most widely favoured to sustain evolutionary change. All living things, or more precisely the units of genetic information from which they are assembled, are envisaged to be subjected to relentless pressure causing them to "struggle for existence". In the harsh glare of exposure to natural forces in a world containing finite resources, those that don't measure up don't survive. Only the most well-adapted and thereby the most competitive will endure.

However, there has always been some unease in understanding how natural selection, in the way that it has just been portrayed as a negative feedback or self-containing process, can be *creative*. How can a "quality control" process be the instigator rather than merely the arbiter of evolutionary change? In fact, it can't be. Rather, it is the ability of living things to generate variation which is primarily creative. What natural selection does, as a quality control process, is to direct and assist the process of change by allowing "fit" (well-adapted) individuals to reproduce, and so increase in number, at the expense of "unfit" (poorly adapted) individuals.

So, the general view is that evolution proceeds by trial and error. Variable units of selection (individuals and genes) are generated, they are exposed to selection pressure (trial), and the misfits (errors) are eliminated in favour of the best fitted. Things, it might seem, can only get better—cruel and wasteful though the means to the end may be!

There is a snag with this view. It is easy to see that the instigation of a quality control system can lead to refinement and specialization in pursuit of a particular goal. But what happens then? Were the process of competitive refinement to continue indefinitely, it is hard to imagine how anything fundamentally new could arise: competitors just get better at doing the same thing. With time the refinements would be expected to involve smaller and smaller adjustments and any significant departures from previous patterns would be selected against in the short term. Ultimately, selection pressure, in the guise of competition, impedes innovation and narrows down choice! The evolutionary process will then get bogged down, generating new details perhaps, but no new themes.

It may be, therefore, that this negative, and at heart analytical view of evolution by natural selection is incomplete. There may be "nurturing" as well as "disciplinarian" processes at work and, as in scientific research, competition may be a hindrance as well as an agent of refinement. Somehow, if major changes are to occur, there has to be some relaxation of competitive pressure, accompanied by a shift in circumstances—an *opportunity* or "selection vacuum"—that provides living systems with the freedom to innovate. In other words, there has to be some evolutionary playtime! Evolution can then proceed more by error (experiment) and trial (does it lead anywhere?) than by the elimination of imprecision.

For a long time, plant and animal breeders have in fact consciously or unconsciously acknowledged the need for evolutionary playtime in order to generate novel varieties. The proliferation of different kinds of dogs and cabbages, for example, has depended on the intervention of human beings in such ways as to give forms that would not have been likely to succeed in wild populations the chance to grow and reproduce. Once freed from the burden of competition, the creativity of living systems can be breathtaking!

In natural populations of organisms, playtime may sometimes be signalled by some external event or change such as a collision with a comet or an alteration in atmospheric conditions. However, it is also vital to appreciate that one of the most fundamental consequences of the dynamic nature of living system boundaries is that these systems create their own new opportunities. Departures from previous patterns can then actively be enhanced by positive feedback as well as constrained: there is carrot as well as stick.

Evolutionary opportunities, selection vacuums, have two components: the emergence of a new context for living and the means to occupy that context. They are also known by another name—perhaps the most challenging abstraction of evolutionary theory and one which highlights the contradictions that can arise from absolute definitions of boundaries—that of niche.

6. Niches—Cradles of Complexity

Niches can perhaps best be understood as open-ended segments of space, time and energy whose boundaries—like the imprint produced by marbles poured onto sand—both define and are defined by the living systems that inhabit them. The ability of a living system to occupy a niche depends on the possession of particular enabling attributes, prescribed partly by genes, and these attributes also define what form the niche will take. For example, an animal with gills can occupy and create a niche in water whilst one with lungs can generate a niche on land.

New niches, new opportunities, arise as a result of feedback interactions amongst both living and non-living components of natural environments. Just as a change in climate can be brought about both by living and non-living agencies and creates new constraints and opportunities, so the evolution of a tree provides an opportunity for a climbing plant to evolve. In such ways, evolutionary processes are self-enhancing or "autocatalytic", building on their past to create new and fundamentally unpredictable opportunities for the future.

The relationship between niche and competition is often obscured by circular argument. On the one hand, it is commonly said that adaptation to new or different niches allows escape from competition because entities with different requirements do not compete. For example, different species of wading birds can co-exist on a mud flat because they have specialized bills that equip them to feed on distinctive kinds of prey. On the other hand, it is argued that niches are never shared because the best-adapted entity would always out-compete less well-adapted ones. So it would seem that adaptation to niche involves both an escape from competition and the means of being competitive!

If such logic is followed to its end points, the inescapable conclusion is that every living entity occupies a unique niche boundary—which is about as illuminating as saying that everyone seated around a dinner table is on a different chair! In the face of such conclusions many evolutionary biologists have been tempted to abandon the concept of niche altogether.

The problem here arises from the tendency to view niches as fully definable goals, that is as discrete "boxes" that are either filled or empty and which exist independently of the evolutionary process. The same kind of thinking explains the presence of particular properties of an organism purely in terms of the *adaptive* value of the genetic information that prescribes them.

Adaptational arguments focus attention on the way the niche boundary determines what the organism is like rather than on how what the organism is like influences the niche boundary. Whilst this might sound like a quibble, it is in fact fundamental to understanding because of the chicken and egg relationship between a dynamic system and its boundary. Adaptational arguments can help explain why some property of a living system *persists* once it has come into being, i.e. because it increases or at least does not reduce survival chances. However, they provide no insights into *how* this property has come into being and pay no heed to the fact that at the time of its appearance any future benefits it could confer might not as yet have been resolved.

Think, for example, of a river system. Told that it was alive, adaptationists could argue that its branching pattern is the result of its ability, honed by natural selection, to respond to the local landscape in such a way as to achieve maximum drainage to the sea. They might not be able to view the pattern as the automatic consequence of the organizational properties of water, as a fluid system finely balanced between association and dissociation, to follow, create and consolidate paths of least resistance.

In fact, as I will explain in later chapters, there *are* living systems that do produce river-like patterns, and these patterns *have* generally been explained primarily in terms of adaptational needs rather than organizational properties. When living systems are viewed as objects, from outside-in, their properties will seem to be *compelled* by external forces. When they are viewed as subjects, from inside-out, their behaviour will seem to be *impelled* by internal forces. The outside-in viewpoint predominates in analytical science and is at the heart of adaptationist thinking about living systems. Viewed from outside-in, a living system will necessarily seem to *respond* to external circumstances and to do so in order to enhance its chances of survival. It must receive signals from the environment and "transduce" them into an internal message that activates an appropriately calculated response based on genetically encoded information. The system will seem like a precision-built assembly of gadgets.

To continue with the riverine metaphor, it may be possible to have more apt ideas about how opportunities arise as the *product* of playful evolutionary processes, by adopting a more dynamic and *open-ended* or *fluid* perspective of niche. As niches first begin to emerge, the full nature of the opportunities that they give rise to may be far from clear. Only later can they be resolved into definable goals that can be attained by adaptive refinement. Then, just when the refinement process looks like reaching its end-point, evolutionary feedback goes and moves the goal mouth and new opportunities start to emerge! Evolution results both from changes in context and changes in genetic information, neither of which can be understood in isolation. *Like any kind of information, genes only make sense in context.*

Here, it may be helpful to reflect again on self-nearself-nonself relationships and what follows when two or more living entities occur within the same space-time-energy boundary (cf. Fig. 1.1). Such co-existence may arise either because the entities are equipped to use the same resources or because one or more of them provides a resource for the other(s).

The consequences of co-existence due to using the same resources are illustrated by the flow diagram in Fig. 1.2. Such co-existence can be due either to divergence from a common ancestor or to convergence from different ancestors—in much the same way that a school can become populated by offspring from the same and different parents.

At first, the entities may co-exist with relatively little mutual interference; all that matters at this stage is that they are able to inhabit the same boundary—no matter how effective they are relative to one another in doing so. However, as they grow or reproduce and the boundary becomes limiting, they will inevitably begin to compete—unless they immediately coalesce. As in a game of rugby, such competition can involve being the first to get a hold of resources ("primary resource capture") and also actively retaining or wresting resources that have already been gained ("combat").

Since there must either be "losers" or persistent conflict, competition inevitably dissipates energy that could otherwise be invested more positively and that can never be fully regained by "winners". The process may also be sustained by what has been described as a "co-evolutionary arms race" or "spiral" in which the competitive ability of the contestants becomes increasingly sophisticated as they put one another under selective pressure. The adversaries have to invest more and more if they are to remain in the same position relative to one another. This situation has been likened to that of the "Red Queen" in "Alice through the Looking Glass" who had to run as fast as she could to stay in the same place.

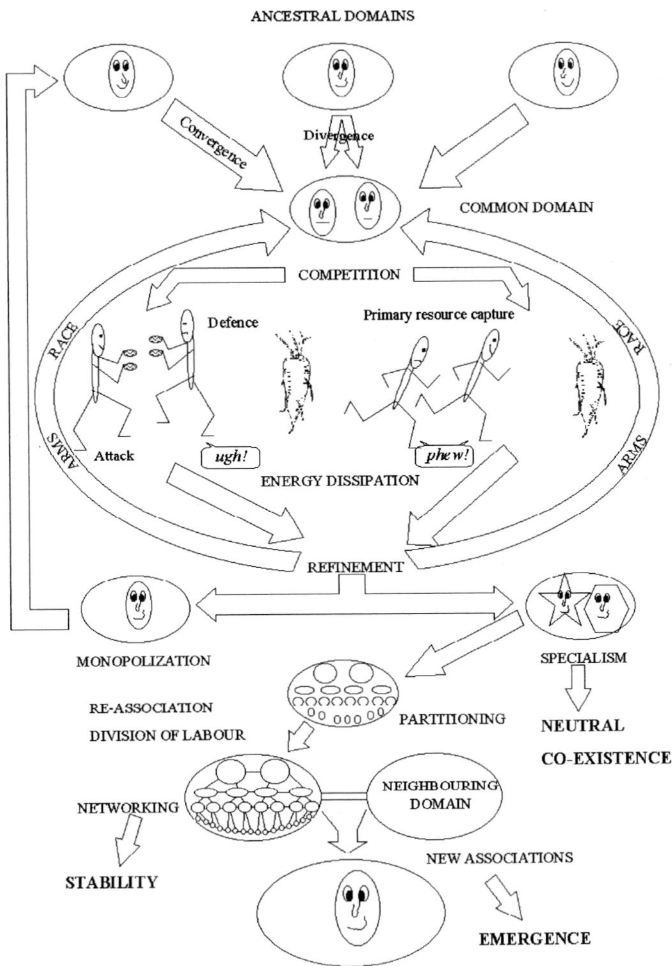

Figure 1.2. The consequences of co-existence due to using the same resources (represented here as a carrot). (Drawn with help from R.E. Guevara)

This refining but dissipative competitive process can be ended in two ways. First, extinction of the opposition may result in the establishment of a monopoly. Second, the refinement of resource capture mechanisms, initially a means of increasing competitiveness, may result in specialization in particular activities—as illustrated by the waders on a mud flat example (see above). Such specialization results in a narrowing down of the domain occupied by individual entities within the original boundary so that they cease to compete directly with one another. In other words, what was once an open field in which the inhabitants could gambol freely becomes progressively closed down and internally partitioned into subdomains. Each subdomain may then in its turn become partitioned, so generating more and more internal detail. Notice here how the apparent paradox that specialization is both the means of becoming competitive and of avoiding competition is resolved by understanding how an initially general domain becomes internally diversified.

The establishment of monopolies and the subdivision of niche boundaries as the end-product of competitive processes depends, however, on the system being or becoming "determinate". This means that the internal partitions between subdomains must remain fully sealed and that the original boundary remains stationary and isolated from all others. However, as the ingredients of an "indeterminate" or open-ended system change, so the original boundary will be prone to shift, opening up prospects of entering unexplored territory and making connections with adjacent boundaries. Connections may also be made between specialisms, enabling them to combine forces. All these events give rise to new opportunities.

In classical evolutionary theory, the process of diversification from a common ancestor into specialized niches is referred to as adaptive radiation. A well-known example is provided by the distinctive species of pouched mammals (marsupials) of Australia. However, it is important to realize that very similar processes, operating over much shorter time scales, result in the formation of cells or tissues with specialized functions in multicellular organisms, and of organisms with specialized roles within societies. A general term which can be applied to all such diversification processes is *differentiation*, whilst the interconnecting processes that supersede them represent *integration*. Integration can result in the formation of assemblies of intercommunicative and ultimately interdependent components or *networks*. When it allows the boundaries of a system to be extended beyond its previous limits, it gives rise to *emergence*.

A rather different route to integration occurs when one entity provides a resource for the other, resulting in temporary or persistent associations. Setting aside any attempt to identify hard and fast categories, the flow diagram shown in Fig. 1.3

attempts to provide an overview of the evolutionary consequences of these associations.

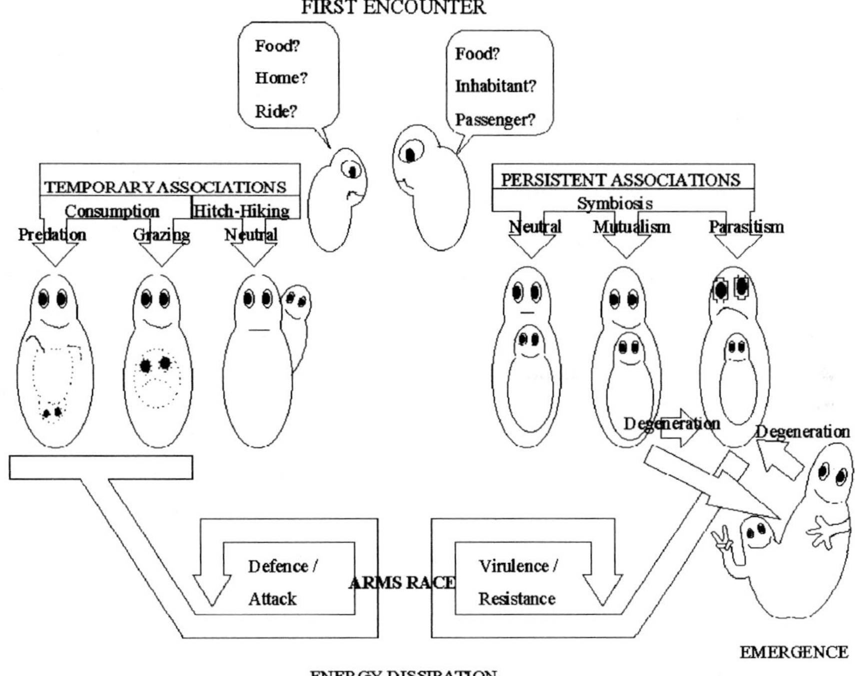

Figure 1.3. The varied relationships which can develop when one living entity provides a resource for another. (Drawn with help from R.E. Guevara)

Temporary associations often involve one entity being consumed by another. This process can be described as "grazing" if the consumed entity is sedentary, or as "predation" if it is mobile.

Persistent associations can all be described as "symbioses", using this term in its original sense as any kind of living together intimately, regardless of the consequences. Traditionally, the concept of symbiosis has been applied only to associations between individual organisms. However, as I will be trying to make clear in later chapters, there are good reasons for extending it to cover all kinds of entities, from genes to societies.

Where one entity draws resources from the other (its "host") without providing any benefits in return, the relationship is parasitic. Parasitism inevitably lessens the chances of survival of the host, so imposing strong selection pressure for the evolution of resistance or rejection mechanisms that seal the host boundary from attack. Such mechanisms in turn impose pressure on the parasite to circumvent them. The resultant alternating selection pressures may therefore precipitate an arms race between host and parasite in the same way as also occurs in temporary associations and between competitors. Where the parasite depends on its host, by reducing the latter's chances of survival it also puts itself at risk. So, at the same time as avoiding resistance mechanisms, there is also a pressure for a dependent parasite to keep damage to its host to a minimum, and indeed to benefit its host in some way.

As has been mentioned, symbioses from which each participant derives benefit can be called mutualistic. In these symbioses, the *partnership* generates a new niche. Over time, the relationship may become so intimate that original boundaries break down or come to define part and parcel of the same entity. Differentiation, allowing complementation between disparate components of a living system may itself depend on the original input of varied genetic information into such ancestral symbioses.

Once a fully integrated union has developed, there may still be possibilities for the original boundaries between the partners to be re-established; apparently self-destructive processes may then ensue, including some which could play a role in ageing.

These possibilities bring the arguments in this chapter full circle, back to the question of what is an individual and what is a self. Thoughout, I have wanted to convey the generality of the arguments—the fact that they can be applied to all kinds of living systems, from molecular to social organizations. I have therefore refrained from saying much, other than by providing examples, about what the participants on the stage of the association-dissociation interplay actually are. It is now time to introduce the players.

CHAPTER 2

SCALING HIERARCHIES: INDIVIDUALS AND COLLECTIVES FROM MOLECULES TO COMMUNITIES

1. Before Beginning

In the last chapter, I suggested that no living thing is quite as discrete as it might at first seem. Rather, there is a dynamic interplay between association and dissociation which recurs across the whole spectrum of life forms, and no individual entity can have complete freedom. Now I want to establish how well this idea relates to what can actually be observed about the way living systems are organized, from the smallest to the largest scales of existence. However, I have an immediate problem.

A fundamental reductionist assumption is that life must have had a definable starting point. This implies that a clear boundary can ultimately be drawn between animate and inanimate processes. However, any ideas about what is and is not living are likely to be prejudiced by how life looks to us at the present time. The components of anything recognisable today as living or as having lived will already have developed such complex interdependencies with one another that their origins and relationships will be impossible to trace. Any explanation of the origin of life in terms of a simple sequence of individual steps will therefore seem to require fantastically improbable co-incidences. On the other hand, if life has evolved by means of complex interactions between associative and dissociative processes, and boundaries are never absolute, the notion of a specific starting point becomes virtually meaningless.

My problem, therefore, is where to begin this chapter? There could be some justification in reaching back to sub-atomic scales, but that would make for a long book! It would also mean falling into the reductionist trap of paying more attention to the exact nature of components than to the general processes that operate within and around living systems. That would mean becoming enmeshed in the need to describe finer and finer detail without ever getting to the root issues, a sorry state which characterizes all too many scientific disciplines!

So, I'm not even going to try to trace the association-dissociation interplay back to the most fundamental components of living systems. Instead, I aim to explore the

...ty of modern-day organisms at organizational scales ranging from molecular to social, and to describe those entities and groupings which come into view at each of these scales. By the end of the chapter, the recurrence of the mainstream theme of association and dissociation should be all too evident. The ramifications of this theme will then be navigated more deeply in subsequent chapters.

2. Molecular Assembly and Disassembly

Within the boundaries of all known life forms, a remarkable interplay occurs between two types of complex chemical compounds. The genetic information specifying the kinds of components from which organisms are assembled is stored and distributed in compounds called "nucleic acids". Compounds called "proteins", on the other hand, provide physical structure and, as "enzymes", accelerate (catalyse) what would otherwise be the extraordinarily slow chemical reactions on which life depends. They also regulate the way that the genetic information is put into effect ("expressed") at different times and places.

The discovery of the structure and mode of assembly of nucleic acids and proteins is widely regarded as one of the major achievements of twentieth century science. The story of this discovery has been told many times, and the flood of incredibly detailed information that has followed in its wake is the subject of innumerable authoritative treatises.

In the next few pages I will therefore only try to pick out the most salient parts of the story and describe how I think that the knowledge that has been gained contributes to our understanding of life patterns.

To begin with, it is interesting to reflect that it was once commonplace to think of a kind of chain of command from nucleic acids to proteins. This view was especially prominent during the euphoria that followed the elucidation of the structure of DNA (see below). It is very much in keeping with a philosophy that sees order being maintained by powerful central administrations that regulate the activities of a labour force.

However, it has since become increasingly clear that the relationship between nucleic acids and proteins is more of a flexible and potentially unstable partnership than a rigid dictatorship. Indeed, the relationship between nucleic acids and proteins is so complex and interdependent that it is difficult, if not impossible to envisage how they could have had separate evolutionary origins.

Here it is worth appreciating that modern day proteins and nucleic acids are themselves collective assemblies. Known technically as "polymers", these compounds consist of component sub-units or "monomers" linked to one another by chemical bonds. The formation and breakage of these bonds is the basis of association-dissociation interplay at molecular levels of organization.

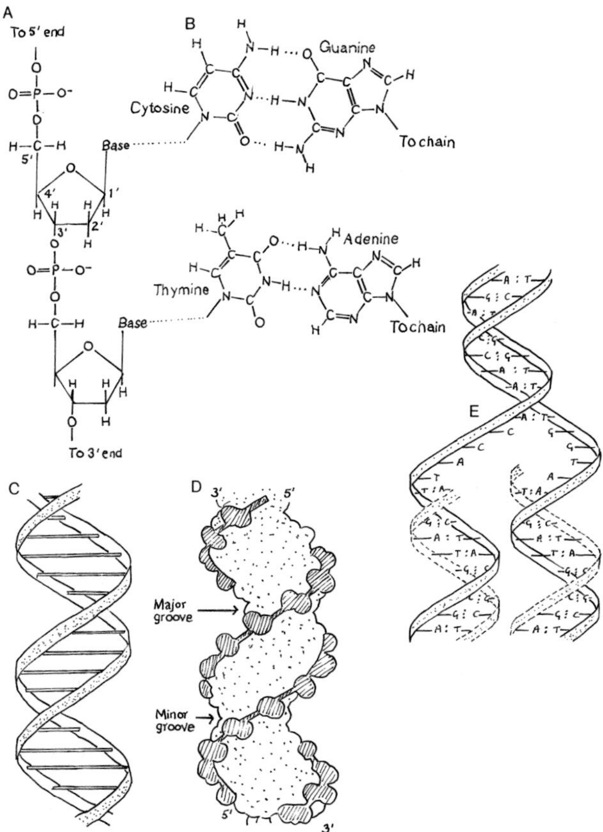

Figure 2.1. The structure and replication of DNA. A, the "sugar-phosphate backbone": note that carbon atoms of the sugar, deoxyribose, are labelled 1' to 5', so giving a definite directionality to the backbone. B, base-pairs joined by hydrogen bonds. C, "ribbon and stick" model, showing how the backbones spiral around each other in opposite directions, linked by base-pairs. D, outline of "space-filling" model, showing compact organization of the molecule and the presence of major and minor grooves—this is characteristic of what is known as the "B form" of DNA; other configurations are possible. E, simple model of replication showing unwinding and assembly of new strands (dotted lines) using the parent molecule as a template.

The monomers of nucleic acids are known as nucleotides (Fig. 2.1). Each nucleotide consists of a sugar (ribose or deoxyribose) linked on one side to a "purine" or "pyrimidine" base and on the other side to a phosphate group (without a phosphate group, the compound is known as a nucleoside). These nucleotides associate in single-file by forming strong, "covalent" bonds. These bonds, which can be thought of as electron partnerships, couple the phosphate group of one nucleotide with the sugar of the next. In deoxyribonucleic acid (DNA), the bases are of four types: adenine, thymine, guanine and cytosine.

DNA molecules are usually double-stranded. That is, they consist of two single-file sequences of nucleotides that spiral around one another in opposite directions. The strands become interlocked by "hydrogen" bonds between complementary bases that fit together rather like jigsaw pieces: adenine with thymine; guanine with cytosine. The hydrogen bonds are due to the slight attraction between molecules that are relatively more positively electrically charged in some parts and negatively charged in others (unlike charges attract, like charges repel). It is interesting to reflect that such a weak (easily broken) association is fundamental to the evolution of life. Not only is it vital, as will be described shortly, to the association and dissociation of DNA molecules, but it is also determines the properties of proteins and that simple but remarkably important substance, water.

The structure of DNA just described is the celebrated "double helix" first proposed by Watson and Crick. The special significance of this structure lies in the fact that exact (or, rather, usually exact) copies of it can be made by a process of replication (Fig. 2.1). This process involves first the dissociation of the two strands along the line of weakness provided by the hydrogen bonds, and then the use of each separate strand as a "template" on which a new partner can be assembled from monomers. Any departures (errors) from the established sequence of bases occurring at or prior to this phase will faithfully (usually) be copied.

Herein lie the roots of heredity which underpin biological evolution's ability to build on its own previous frameworks rather than constantly having to start afresh. The sequence of bases along the DNA molecules spells out the "genetic code". This code provides an effective means for storing and retrieving information needed to mould (or remould) life forms from basic ingredients, in a kind of reproducible "shorthand". The exactness of the replication process prevents all the hard-won refinements achieved through past evolutionary trial and error from being abandoned with each new generation.

At the same time, there are many means by which the information encoded within DNA can be changed. These can involve the addition, elimination or substitution of bases at particular sites (known as "point mutation") or various kinds of rearrangements and recombinations of nucleotide sequences (see Chapter 4). They provide a capacity for error which at least partially frees living systems from the constraints of total adherence to their genetic past. They allow room for experiment (error and trial) and the creation and exploitation of evolutionary opportunities.

It is remarkable that whereas the roman alphabet used to convey the English language requires twenty-six characters, the DNA alphabet needs only four characters for all the biological diversity it specifies. This diversity partly comes from using not just each single base as an information unit, but groups of three bases, "triplet codons", in this role. Even so, it would be a dull language which had only sixty four words (i.e. the total possible number of different sequences of three, 4^3).

In fact, the full richness of the genetic language is only brought out when the four-letter DNA alphabet is expanded into the twenty-letter alphabet of a potentially infinite variety of proteins. Proteins are polymers of amino acid monomers (Fig. 2.2). There are twenty different common types of amino acids (rarer amino acids are modified versions of the common types). The amino acids become linked together by covalent "peptide bonds" in a linear sequence several tens to several thousand monomers long, so forming a "polypeptide". Some proteins consist of single polypeptide chains, whilst others consist of associations of two or more polypeptide sub-units.

There are four organizational levels at which the components of proteins interact. The so-called "primary structure" is the actual linear sequence of amino acid side chains, which stems from a "backbone" of carbon atoms linked by peptide bonds. This backbone is not rigid, but can be deformed—twisted, bent or pleated—into a variety of configurations or "secondary structures", maintained by hydrogen bonds. The twisted, bent or pleated structure can further be folded upon itself to produce the three-dimensional conformation of the whole polypeptide. This tertiary structure is in turn determined by spatial constraints and the formation of covalent and hydrogen bonds which cross-link the molecule. Finally, the quaternary structure is due to the pattern of association of separate polypeptide units which make up composite or "multimeric" proteins.

The secondary, tertiary and quaternary structures together determine the boundary configurations of proteins. These configurations in turn define the "recognition" sites where proteins can associate, temporarily or persistently, with other molecules in the manner of sophisticated, somewhat flexible, locks and keys.

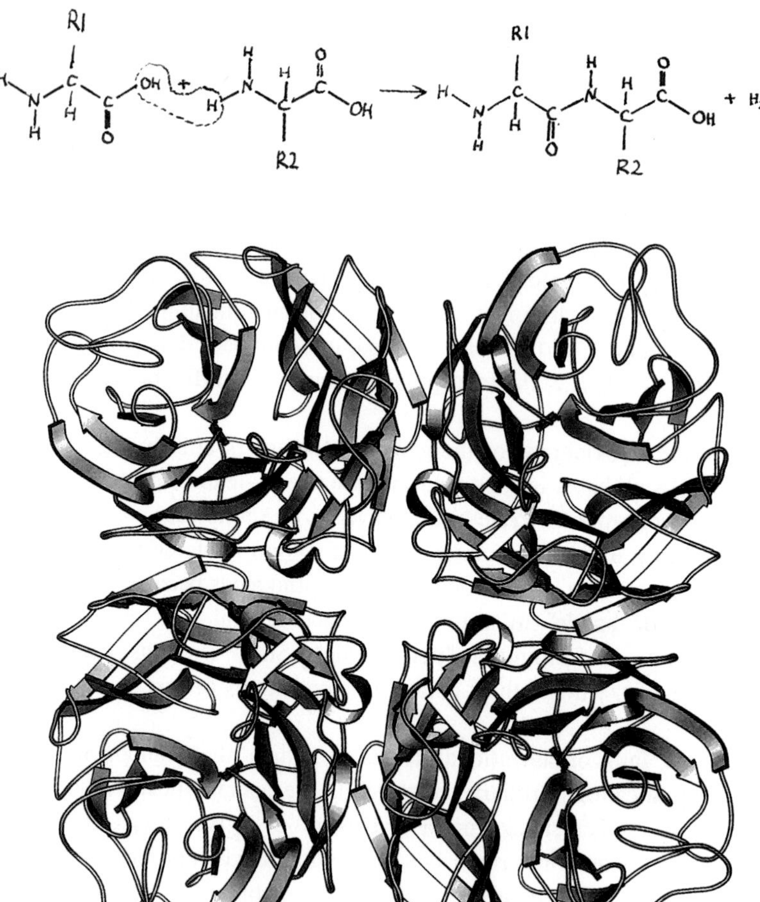

Figure 2.2. The structure of proteins. Top, formation of a peptide bond through removal of water between adjacent amino acids. Below, schematic diagram of secondary, tertiary and quaternary structure of the enzyme, neuraminidase, from the influenza virus. (Courtesy of G.L. Taylor)

Only the primary structure of polypeptide chains is directly specified by the genetic information in nucleic acids. The secondary, tertiary and quaternary structures represent "emergent properties" of the polypeptide chains themselves. Sometimes more than one configuration is possible given a particular primary structure. There is also evidence that the polypeptide chains can undergo further changes, known as "processing", following their initial synthesis from messenger RNA (see below).

In short, the complexity of protein molecules illustrates the ability of associative processes repeated on different hierarchical scales to create a diversity of forms using a simple initial input of information—the four kinds of bases in DNA.

Proteins are products rather than generators of genetic information. This is because, in accord with what has been called the "central dogma" of molecular biology, the conversion of genetic information from DNA into proteins is both indirect and one-directional. That is, proteins are only assembled by means of the "transcription" (see below) of DNA into intermediary ribonucleic acids (RNAs), and the twenty letter alphabet of proteins cannot be recondensed into the four-letter alphabet of DNA.

However, there are several ways in which proteins can influence their own production and properties. Firstly, as enzymes, proteins catalyse the process of replication which is the very basis of heredity. This process, although easily envisaged from the double helix model (Fig. 2.1) in fact requires quite a complex sequence of operations involving more than twenty different proteins. To give you a flavour of the details (gloss over the remainder of this and the next paragraph if you don't like the taste!), the replication process in the bacterium, *Escherichia coli*, occurs as follows. First, a DNA helicase enzyme (enzymes are referred to by adding "ase" to the structure and/or process on which they operate) binds to a site known as an "origin of replication" and begins to untwist the helix. The untwisting breaks the hydrogen bonds between complementary bases and produces "supercoils", similar to those which form when separating the strands of a rope. The supercoils are cancelled out by the action of a DNA gyrase enzyme which inserts twists in the opposite direction. Energy, provided chemically by removal of a phosphate group from molecules of the nucleoside, adenosine triphosphate (ATP), is used up during the action of both DNA helicase and DNA gyrase. Once the strands have been separated, they are kept apart by a composite of six or seven proteins called a "primosome", which binds to the DNA. RNA polymerase then binds to sites known as "promoters" (see below) and synthesizes two short molecules of RNA, with bases complementary to each of the two DNA strands. These molecules are required as "primers" for synthesis of DNA by DNA polymerases which are unable to bind directly to the DNA templates.

DNA polymerases can only synthesize new strands of DNA in one direction. This implies that DNA replication can only be continuous (in fact using the enzyme DNA polymerase III) on one of the original strands; on the other strand it is achieved discontinuously. Discontinuous synthesis requires in turn: another RNA primer-producing molecule, RNA primase; a DNA ligase which can link separate DNA links together; DNA polymerase I, which can excise the RNA primer molecules and replace them with DNA. The latter two enzymes also play an important role in "proof reading" and corrective "repair" of DNA containing inaccurately base-paired nucleotides. These processes are essential for the maintenance of an unchanging genetic message.

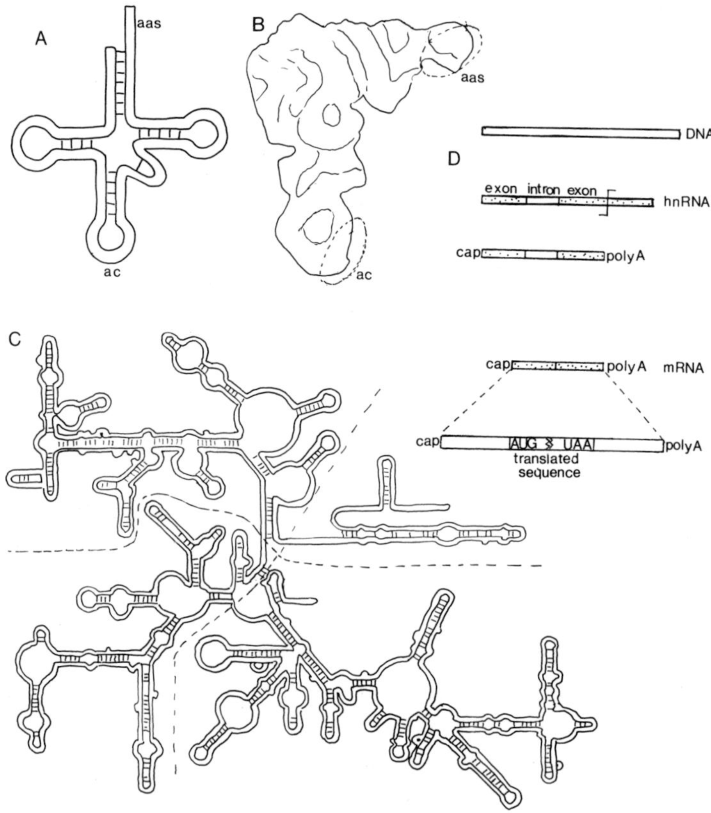

Figure 2.3. Three kinds of RNA. A, general, "cloverleaf" secondary structure of a tRNA molecule, showing position of hydrogen bonds between base pairs, anticodon (ac) which attaches to mRNA and amino acid acceptor stem (aas). B, outline of tertiary structure of tRNA. C, secondary structure of prokaryote 16S rRNA, showing four possible folding regions. D, eukaryote mRNA and its origin from DNA by means of "processing" (capping, tailing and splicing) of the primary transcript, hnRNA.

Returning to the complex relationships between proteins and nucleic acids, the story continues when the inherited genetic information within DNA is converted or "transcribed" into ribonucleic acid (RNA). RNA differs from DNA in that its sugar component is ribose rather than deoxyribose, and uracil replaces thymine as the pyrimidine base complementary to adenine. RNA is synthesized using one or other, or exceptionally both, of the complementary DNA strands as a template to which RNA polymerase enzymes bind, initially at specific sites known as promoters. Whether or not the RNA polymerases bind to the promoter sites and proceed to assist the synthesis of RNA may in turn be determined by proteins which bind to the DNA in the promoter region. Binding of RNA polymerases to a promoter is also influenced by the configuration of their polypeptide components. In bacteria, RNA polymerases are in fact complex "holoenzymes", comprising a "core" enzyme functioning in RNA synthesis itself, and a "sigma" factor. The latter controls binding to specific promoters, which may in turn be of several different kinds.

There are three main kinds of RNA (Fig. 2.3). Messenger RNAs (mRNAs) contain the information from DNA in a translatable form. Transfer RNAs (tRNAs) specifically bind individual amino acids and attach to corresponding triplets of bases on mRNA. Ribosomal RNAs (rRNAs) are contained within specialized subcellular entities (ribosomes) where tRNAs are assembled onto mRNA, allowing the amino acids they carry to be aligned and linked together into polypeptide chains.

In non-bacterial organisms these different types of RNA are transcribed from DNA by three distinctive RNA polymerase enzymes. RNA polymerase I produces most of the ribosomal RNA. RNA polymerase II is responsible for production of mRNA. RNA polymerase III transcribes tRNA and some rRNA. All these enzymes contain from six to ten polypeptide subunits.

Transcription is not, however, the end of the story of RNA production because, with the exception of most bacterial rRNA, the first products of transcription require modification or "RNA-processing" by further enzyme systems before they can function. Of particular interest here is the mRNA of "eukaryotes"—animals, plants and fungi. The RNA transcript which is to become mRNA in these organisms is often longer—sometimes much longer—than the final, functional product. This is because one or more sometimes large portions of the DNA, known as "introns", do not code for parts of functional proteins. However, they are transcribed into what is known as heterogeneous nuclear RNA (hnRNA) and have to be removed by a process of RNA splicing (Fig. 2.3). This process cuts out the introns and joins together the remaining coding regions or "exons". In addition to splicing, both ends of the precursor mRNA are modified. One end, corresponding to the beginning of the message, is "capped" enzymatically by addition of a modified (methylated) guanine nucleotide. The other is "tailed" by the addition of up to two hundred adenine nucleotides (a "poly A tail").

During protein synthesis, each amino acid is activated with the energy-rich molecule, ATP (see above), and then coupled to the appropriate tRNA molecule by a highly specific enzyme, amino-acyl-tRNA synthetase. The resultant aminoacyl tRNA then attaches to the appropriate triplet sequence in mRNA by means of its own complementary triplet sequence or "anticodon". This process occurs at "ribosomes" (see below), which work along the mRNA, translating the sequence of triplet codes into the sequence of amino acids which provides the primary structure of proteins.

The role of proteins in these fundamental processes of transcription and translation which are key to their own production raises a profound problem. Since proteins contain at least a few tens of amino acids, requiring at least a hundred or so nucleotide units of genetic information in DNA, how can one class of molecule exist without the other? The problem is made all the more difficult because although nucleic acids may be capable of replicating very slowly on their own, they do so relatively inaccurately, so that the precision of their genetic message would be quickly broken down. Perhaps, as was hinted before beginning, the problem is best resolved not by trying to analyse complex evolutionary end-products, but by recognizing the interactive feedbacks that can generate complex interdependencies from simple inputs.

3. Assembling molecular assemblies—ribosomes and chromosomes

As if things weren't already complex enough, the story of interdependency between proteins and nucleic acids is still far from complete! This is because these compounds can become physically integrated, by means other than chemical bonds between atoms, to produce higher-order composites known as nucleoproteins.

One kind of nucleoprotein has already been mentioned, the ribosome. Ribosomes are similarly constructed in both eukaryotes and bacteria, although they are slightly smaller in the latter. They are composed of two subunits, of which the smaller binds to mRNA and the larger contains the sites for locating aminoacyl tRNAs.

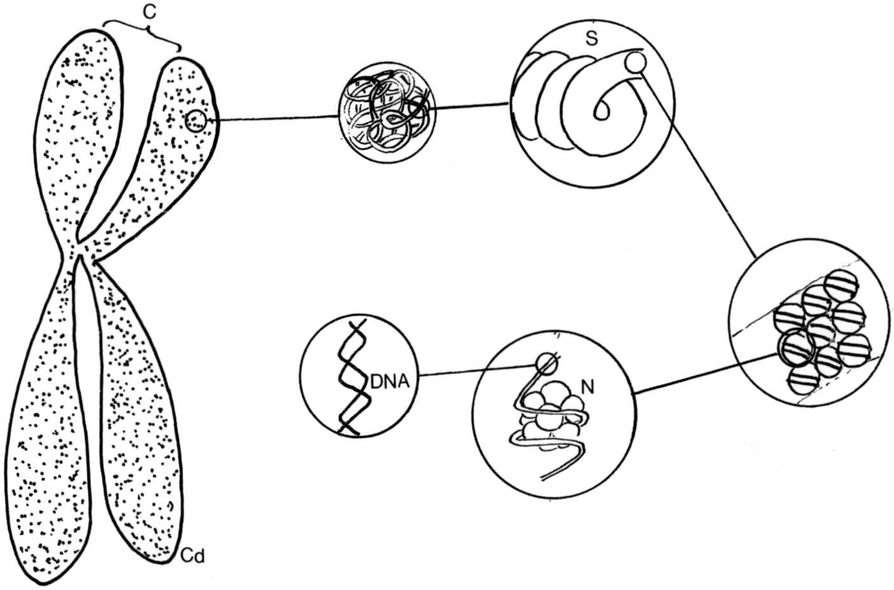

Figure 2.4. The structure of eukaryote chromosomes. C, chromosome, as seen at "metaphase" (see Chapter 4), after replication and immediately prior to cell division; S, solenoid; N, nucleosome consisting of an octet of histone proteins entwined in DNA.

In bacteria, the large subunit of ribosomes has been shown to contain thirty four proteins and a large and a small rRNA molecule. The small subunit contains one large rRNA molecule and twenty one proteins. It was shown as early as 1968 that the small subunit can assemble itself when its components are mixed under appropriate conditions within a test tube. This self-assembly is an ordered process in which it is necessary for some proteins to attach first to the RNA, before others can follow. Another interesting feature of ribosomes is that rRNAs have been shown to have very similar secondary structures throughout bacteria and eukaryotes, even though the actual primary sequence of nucleotides within them may differ considerably.

Fully extended, most DNA molecules within cells are extremely long. This raises the question as to how they can be compactly packaged. In bacteria much of the genetic information is contained in a single DNA molecule, or "chromosome". Stretched out this molecule would be more than 1 mm long, about a thousand times longer than the average bacterium. In this case, the DNA is joined end-to-end to form a circle. Extra twists occur in this circular molecule causing it to coil up on itself or "supercoil"—just as an elastic band does if it is twiddled. The supercoiled structure is further complexed with RNA and protein to give a compact arrangement. As well as the chromosome, other circular DNA molecules known as plasmids are often found in bacteria; their significance will be discussed in later chapters.

In eukaryotes the DNA is arranged in linear rather than circular chromosomes. In a single cell there may be many more than one chromosome, each made up of an elaborate nucleoprotein complex known as "chromatin". In chromatin, each two turns of the supercoiled DNA molecule encircle an ellipsoid body made up of two sets of four "histone" protein molecules, so forming an assembly known as a "nucleosome". The nucleosomes are in turn joined together in a helical array or "solenoid" (Fig. 2.4). Life's "mortal coil" might better be thought of as coils of coils!

4. Genes and Supergenes

The organisation of DNA within the linear or circular arrays known as chromosomes has important implications for the patterns in which genetic information is distributed to subsequent generations when an organism reproduces. Here the important point to grasp is that the DNA in a chromosome contains a series of sequences that are referred to individually as genes. These sequences are commonly a thousand or more base pairs long, and indeed can be much longer if they contain introns (see above).

Apart from those which code for rRNA and tRNA, the majority of genes are believed to code for a single polypeptide. In order to function in this way, it is important that each gene is deciphered in a very precise way, with definite starting and ending points—otherwise the message it contains would literally be nonsense.

So, at last it seems that there might after all be a basic, discrete unit of biological organization, give or take a few introns, that natural selection can act upon unambiguously—the gene! It may not be possible because of the problems of boundary definition to define individuals. However, in the gene surely there is something *particulate*, something with absolute boundaries whose passage through the generations can be plotted with mathematical precision.

Not so fast—life is never quite as simple as that! First appearances turn out to be deceptive, and the boundaries of genes are in fact no more clearly definable in the long term than are those of individuals. Once again, dissociative and associative processes interact, providing the basis for differentiation and interdependency in an evolutionarily creative way that defies rigid conservatism.

To begin with, there is some evidence that particular regions of DNA are capable of giving rise to more than one product. This can be due either to overlapping translatable information or varied processing.

More generally, any genes that are "linked", i.e. which occur together in the same chromosome, can be considered to be parts of the same parcel and will tend to be transmitted *en bloc* to subsequent generations. Linked genes cannot be re-assorted into varied combinations with one another (i.e. "recombined") as readily as can genes occurring on separate chromosomes. There are processes which can cause recombination even of genes on the same chromosome (see Chapter 4). However, the closer together—the more closely linked—genes are, the less likely they are to dissociate.

There are, therefore, variable degrees of association between genes. At one extreme are those genes on different chromosomes that are randomly distributed into subsequent generations and so have an equal chance of occurring or not occurring together in the same individual. At the other extreme are those that are closely linked and so assort non-randomly (such genes are sometimes described as being in "linkage disequilibrium").

For several reasons, very close linkage can be both a cause and a consequence of interdependency between genes, resulting in the formation of so-called "supergenes" or "co-adapted gene complexes". In tune with the discussion of niches in Chapter 1, closely linked genes can have either a common origin, arising from an ancestral gene that has duplicated and diversified, or a disparate origin. Interdependence due to common origin may involve the formation of multimeric proteins containing different polypeptide sub-units, such as the red blood cell pigment, haemoglobin. Alternatively, it may be due to the formation of families of similar but distinctive proteins such as antibodies. Interdependence of genes having a disparate origin occurs when the genes code for complementary functions.

As well as being physically linked, the components of supergenes are also likely to be expressed at the same time as one another, due to being subject to the same regulatory controls or "on-off switches". Such controls are needed because as an organism's circumstances change, it is not appropriate to express all the information encoded in its genes all the time: to do so would result in inefficiency and interference between contrasting functions (see Chapter 4). A classical example is the "lactose operon" in the bacterium, *Escherichia coli* (Fig. 2.5). Here transcription units specifying distinctive enzymes involved in the assimilation of the sugar, lactose, by the cell are all governed by the same promoter. Binding of RNA polymerase to the promoter depends on whether or not a repressor protein occupies an adjacent, slightly overlapping, "operator" site and also on whether an activator protein complex binds next to the promoter. Each transcription unit might be thought of as a separate gene, but in order to be functional it depends, like its neighbours, on the situation at an additional site.

It is therefore too simple to regard genes as absolutely definable, independent bits of information that are individually and separately exposed to natural selection. They are parts of parcels and parcels of parts. Also, if the information that they contain is to be expressed, they need to be located within some kind of arena, some context, that isolates them from the external environment.

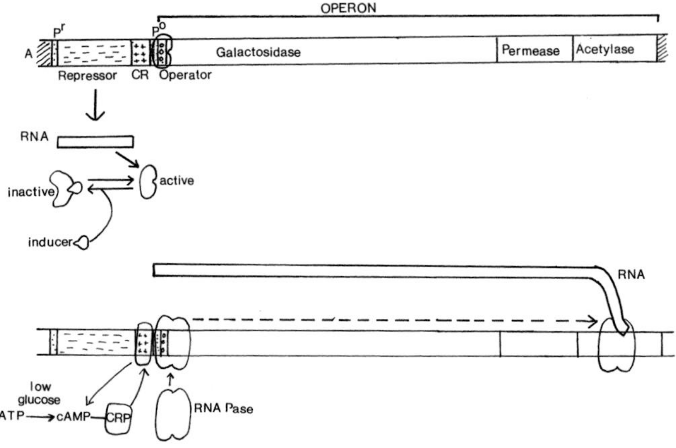

Figure 2.5. Diagram illustrating the structure and regulation of the "lactose operon" of the bacterium, *Escherichia coli*. In the absence of an "inducer" (allolactose), an active repressor protein is produced by a regulatory gene with its own promoter (P^r). This repressor protein binds to an "operator" site, so blocking the action of RNA polymerase (RNA Pase). In the absence of active repressor and in the presence of a cyclic adenosine monophosphate (cAMP)-cAMP receptor protein (CRP), transcription, through their common promoter (P^o), of the operon genes encoding enzymes involved in lactose assimilation is activated. When associated with cAMP, the boundary properties of CRP are changed, so enhancing its ability to bind to a site (CR) adjacent to the operon and stimulate RNA polymerase action.

5. Becoming Insulated—Forming Cell Boundaries

The inadequacy of nucleic acids and proteins alone as sustainable living systems is evident from the behaviour of viruses. Viruses are nucleoproteins that are completely inactive when in the external environment, where they are no more "alive" than grains of salt. However, they become able to generate more of themselves, if they can gain access to the interior of truly living "host" organisms (Fig. 2.6).

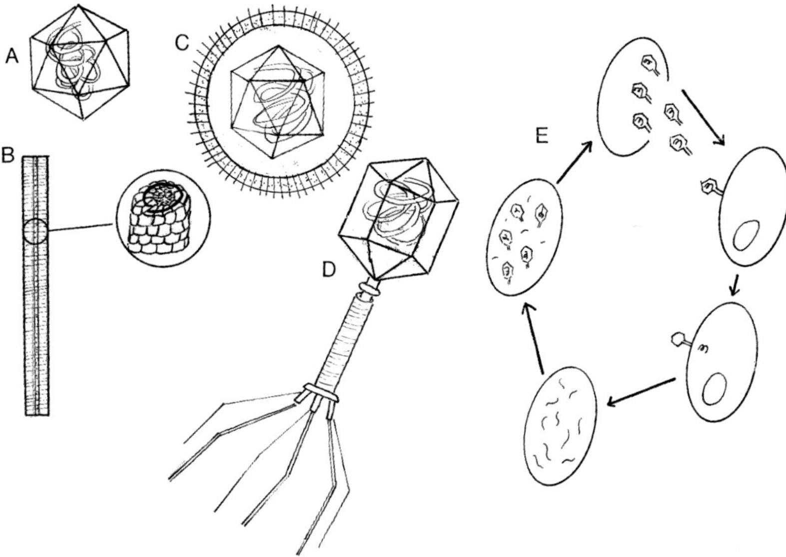

Figure 2.6. The structure and proliferation of viruses. A,B, two basic forms, one with a polyhedral, typically icosahedral (20-sided) protein coat ("capsid") around the nucleic acid, the other with protein subunits spirally wound around the nucleic acid (as in tobacco mosaic virus). C, an enveloped form, surrounded by a lipoprotein capsule. D, complex form of "T-even bacteriophages". E, typical "lytic" (cell-destroying) cycle in which the virus nucleic acid is replicated and translated within the host cell, allowing new virus particles to emerge and spread infection.

What enables true organisms both to be alive and to serve as hosts for viruses is the presence of a watery zone between the genetic store and the outside world. This zone provides the medium in which the energy conversions necessary to sustain a dynamic existence can take place. It has an external boundary, so defining systems known as cells, and is therefore referred to as "cytoplasm" (i.e. cell fluid). Both the bounding of cytoplasm, and the energy conversions within it, involve two more fundamental classes of organic compounds, the water-excluding "lipids" and the water-incorporating "carbohydrates".

The complexity of cytoplasmic boundaries varies. There is an especially fundamental distinction between the "prokaryotic" cells of bacteria and the "eukaryotic" cells of plants animals and fungi (Fig. 2.7).

In prokaryotic cells there is a cell wall (except in organisms known as mycoplasmas) and plasma membrane around the cytoplasm. However, there are no internal partitions between the genetic information and cytoplasm or between different cytoplasmic components. There is therefore relatively little scope for differentiation of specialized functions.

Eukaryotic cells are, by contrast, markedly subdivided internally into cytoplasmic regions and components that serve specialized functions. As the term eukaryotic implies, the chromosomes are contained in a true nucleus which is partitioned from the cytoplasm by a double layer of membranes. There are membrane-bound "organelles" such as "mitochondria", "chloroplasts", "Golgi bodies", "lysosomes" and "peroxisomes". There may be one or more solution-filled "vacuoles", and there are membrane networks known as "endoplasmic reticulum" (ER) which ramify through the cytoplasm. Ribosomes are assembled in a distinct region of the nucleus, the nucleolus, and are exported to the cytoplasm where they either remain free in the cell sap ("cytosol") or attach to the ER. Within the cytosol may be "cytoskeletal" proteins, "microfilaments", "microtubules" and "microtrabeculae", which provide for support and movement within the cell. DNA is contained not only in the chromosomes of the nucleus, which are collectively referred to as the nuclear "genome", but also in some organelles, notably mitochondria and chloroplasts. Mitochondria are the sites where chemical energy in the form of ATP is generated by means of oxygen-consuming (aerobic) respiration. Chloroplasts are the sites in plant cells where carbon dioxide and water interact, powered by light energy, to produce carbohydrates and oxygen during the process known as photosynthesis. The presence of DNA, as well as ribosomes, in both these kinds of organelles has been taken to suggest that they may originally have been free-living prokaryotes that became integrated by enclosure within another cell boundary to form an "endosymbiosis" (see Chapter 1). Eukaryotic cells are therefore veritable parcels of parcels, externally bounded and internally partitioned by membranes.

It is very important to appreciate that as a consequence of their composition and architecture, cell membranes constrain but do not prevent the flow of information and resources into, out of and within cells. As is vital to a dynamic system, they are partial boundaries which serve to maintain a balance between the opening up and sealing off of communication channels.

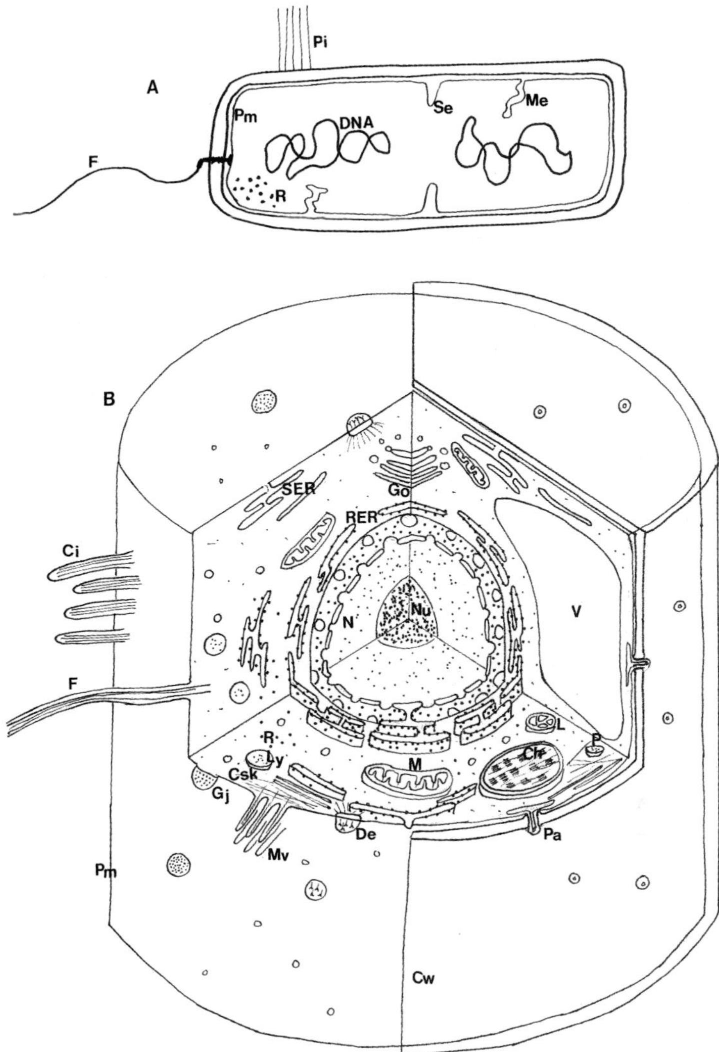

Figure 2.7. Diagrams illustrating the general structure of prokaryotic(A) and eukaryotic(B) cells as seen in a longitudinal slice and a three-dimensional cut-away respectively. In B, features of animal cells are shown on the left of the diagram whilst plant cell features are shown on the right. Ch, chloroplast; Ci, cilia; Csk, cytoskeletal elements (microfilaments and microtubules); Cw, cell wall; De, desmosome; F, flagellum; Gj, gap junction; Go, golgi body; L, leucoplast; Ly, lysosome; M, mitochondrion; Mv, microvilli; Me, mesosome; N, nucleus; Nu, nucleolus; Pa, plasmodesma; Pi, pilus; Pm, plasma membrane; R, ribosomes; RER, rough endoplasmic reticulum (with ribosomes attached); SER, smooth endoplasmic reticulum; Se, septum developing in a dividing cell.

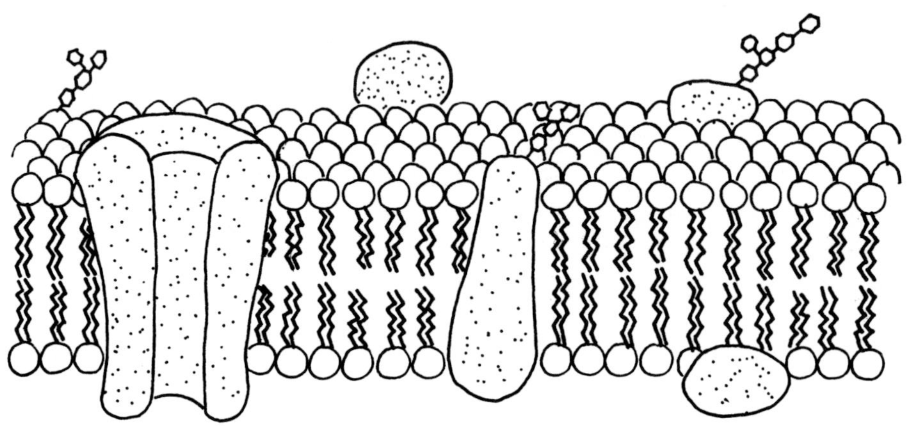

Figure 2.8. The general structure of cell membranes. In addition to the phospholipid bilayer are various proteins (stippled), including channel proteins (left) and carbohydrate chains attached either to proteins (hence forming "glycoproteins") or to lipids (forming "glycolipids").

The principal ingredients of all cell membranes are molecules of "phospholipids" (Fig. 2.8). These elongated molecules are attracted to water (i.e. "hydrophilic") at one end and repelled by water (i.e. "hydrophobic") elsewhere. When placed in contact with water they therefore tend to line up, with the hydrophilic "heads" facing outwards and the hydrophobic "tails" facing inwards, so forming a self-sealing double layer ("bilayer"). This layer is "semi-permeable": it allows passage of water, gases such as oxygen and carbon dioxide and small, relatively hydrophobic molecules such as ethanol, but is impermeable to most water-soluble molecules.

Also present in cell membranes are a variety of proteins, some of which provide channels through which particular kinds of substances are transported. Certain of these proteins can be thought of as pumps, which directly or indirectly use energy derived from ATP to drive solutes across membranes by a process of "active transport". Some of these solutes consist of electrically charged atoms or groups of atoms (i.e. "ions").

Because of active transport, cells can maintain concentrations of solutes, including ions, within their interior that differ from those in their external environment. The cells therefore develop a difference in electrical charge, i.e. a "potential difference" known as the "membrane potential" and are able to operate as "energy sinks" into which resources flow from outside. The maintenance of this membrane potential is essential to a cell's ability to function and stay alive, and it has been estimated that as much as a third of an animal's energy expenditure serves this need.

The presence of higher internal than external solute concentrations also leads to a tendency for water to enter cells by a process of "osmosis" which, unless counteracted in some way, will cause the cell to expand. Osmosis can only be maintained as long as the cell remains capable of generating and using internal energy. Otherwise diffusion (transfer of solutes from high to low concentration solutions) takes over and solute concentrations inside and outside the cell become equal.

The use of internal energy conversions to achieve a net gain in energy from its environment can be understood on the same lines as an engine which is able to run indefinitely as long as it is supplied with the fuel which drives its continued uptake of fuel. However, as with the engine, there is still a need for some external assistance to get the system started.

Some means of starting off with an internal supply of "fuel" could therefore be fundamental to the evolution of life. For example, if a lipid membrane were to form in a salty puddle, and so enclose within itself a concentrated solution, the resulting membrane-bound system would be able to draw water from its surroundings immediately following a rain shower. Providing some mechanism could be set in place that could sustain and amplify this ability, then the self-regenerating processes of life could be set in train. Ultimately this mechanism would involve the uptake or generation of reduced forms of carbon (i.e. carbon compounds containing hydrogen) and the integration of proteins within the membrane.

Besides the processes of diffusion and active transport which occur across intact membranes, there are other modes of entry and exit which depend on openings or rearrangements in the membrane. An example of the former are the pores in the double membrane which surrounds the nucleus (Fig. 2.7). These pores are formed where the inner and outer nuclear membranes merge, so bringing the nucleus into communication with the cytosol. The outer nuclear membrane also has openings where it merges with the rough ER.

Rearrangements of external and internal membranes allow the enclosure of imported materials by means of infolding ("invagination") or enfolding ("engulfment"), or the exclosure of exported materials by coalescence (Fig. 2.9). All of these processes depend on the ability of membranes to break and reseal (i.e. to dissociate and re-associate). Where invagination and engulfment occurs in plasma membranes, they are referred to as "endocytosis" and "phagocytosis" respectively. Phagocytosis allows large particles to be ingested. It may often involve specific association between molecules known as "ligands" on the surface of the particle and "receptors" made of "glycoprotein" (carbohydrate-protein complexes) on the surface of the membrane. Endocytosis may also involve receptors, or it may be relatively non-specific, resulting in the enclosure of small amounts of fluid present at the cell surface by a process of "pinocytosis".

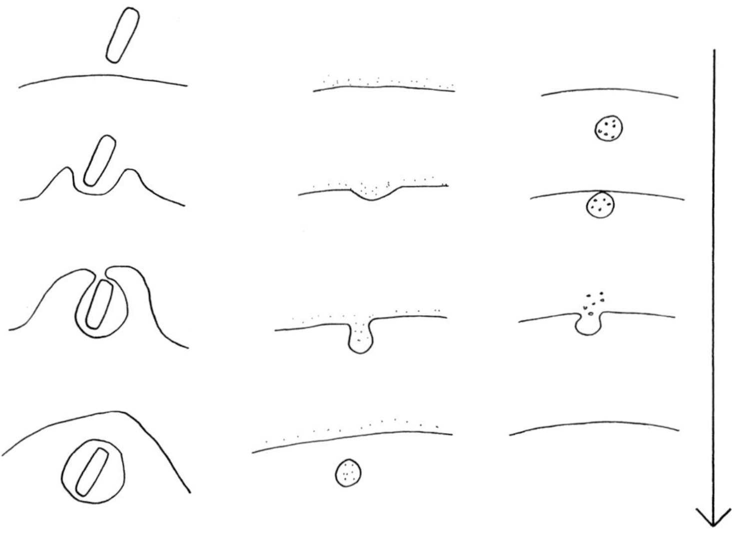

Figure 2.9. Modes of entry to and exit from cells involving disruption and coalescence of membranes. The arrow represents time. Left series, phagocytosis; centre, endocytosis; right, exocytosis.

Phagocytosis and endocytosis enable animal cells to ingest items, but are precluded by the presence of cell walls in plants, fungi and bacteria. These latter organisms therefore have to obtain resources from their external environment purely by *absorption* of water and light or chemical energy, a fact which is fundamental to their patterns of evolution.

Production of cell walls is one way of counteracting the influx of water by osmosis that occurs in "hypotonic" environments (i.e. where external solute concentration is lower than internal concentration), in that it physically resists the outward expansion of cells. Animal cells either avoid the problem by inhabiting "isotonic" environments or have special devices to pump out the excess.

The resistance of cell walls to expansion in hypotonic environments results in a build up of internal pressure ("turgor") equal to the difference between the external and internal water potential. In much the same way as a balloon, the pressurized cells take up a shape which depends on the relative deformability of the cell wall over all parts of its surface. Outward form is therefore dictated by the potentially changeable counteraction of expansive and constraining processes at cell boundaries, rather than being anchored to some form of persistent internal skeleton. The possession of a deformable cell wall therefore combines ordering self-constraint with the freedom to explore environments in a diversity of guises.

It therefore follows that the strength properties of cell walls are critical to their form and function. In bacteria, strength is provided by a meshwork of chemically cross-linked sugars and peptides. The cell walls of plants and fungi are "microfibrillar", containing elongated fibres that confer tensile strength, embedded in a matrix that cements these fibres together in a composite analogous to fibreglass. In plants, the fibres are composed of the carbohydrate, cellulose. This is a remarkable linear polymer in which chains of glucose monomers align with one another to produce fibres referred to as "micelles", "microfibrils" and "macrofibrils" according to the scale of organization. In fungal cell walls, the microfibrillar component is usually "chitin", a polymer of "N-acetyl glucosamine" monomers which is also present in the outer coverings of a variety of multicellular invertebrate animals.

It is not, however, just the strength properties of cell walls that is important to their role in defining cell boundaries. As with membranes, their permeability, or resistance to passage of materials is also critical, particularly in environments where there are possibilities of losing resources or taking in toxins. Here there are advantages in being able to seal the walls with hydrophobic materials. These materials include suberin (the main component of cork) and lignin (a component of woody tissues) in plants and various polymers of phenolic compounds and hydrophobic polypeptides known as "hydrophobins" in fungi.

6. Gathering Cells

Some organisms live out most of their lives as single cells—as the earliest life forms to evolve on the planet are usually assumed to have done. The degree of sophistication that has been achieved by some eukaryotic single-celled organisms, known as "protists", is quite remarkable. A good example is *Paramecium* (Fig. 2.10) which has structures analogous to the mouth, throat, anus and kidneys of a mammal. Another protist, *Euglena*, illustrates the artificiality of any absolute boundary definition between animals and plants by being capable of ingestion, locomotion and photosynthesis. The high levels of individual sophistication found in protists are generally associated with, if not vital to, a relatively independent life-style. However, such self-sufficiency may actually tend to impede co-operative interactions—a theme which is repeated at all life's organizational scales.

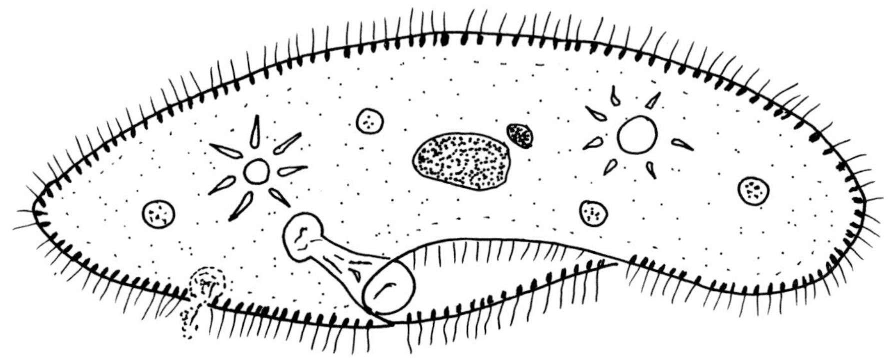

Figure 2.10. *Paramecium*, a sophisticated single-celled organism.

However successful unicellular organisms might be, like all other life forms they have to be able to "proliferate"—to make more of themselves, if they are not to die out (see Chapter 3). In many cases, the means of proliferation involves the expansion and division of cells to make more cells. If the cells dissociate fully from one another following such divisions, because they are free either to drift or propel themselves apart, then they will retain a relatively independent existence.

There are, however, many reasons why cells may not have the freedom to disperse from one another. They may be limited by the boundaries or surfaces within or upon which they grow. They may be attracted to one another and/or adhere when they come into contact. They may not separate from one another following division, or the division process itself may be incomplete.

The ability of cells to proliferate therefore automatically implies that they are liable to associate and so produce collective assemblies. At the simplest levels of organization, as in yeasts and bacteria, these assemblies may consist of gatherings or "colonies" of identical cells, with each cell being fully capable of a free-living existence. At more complex levels, increasing specialism and consequent division of labour and interdependence culminate in the evolution of truly multicellular organisms. In the process, the mantle of individuality is passed almost imperceptibly from cell to collective.

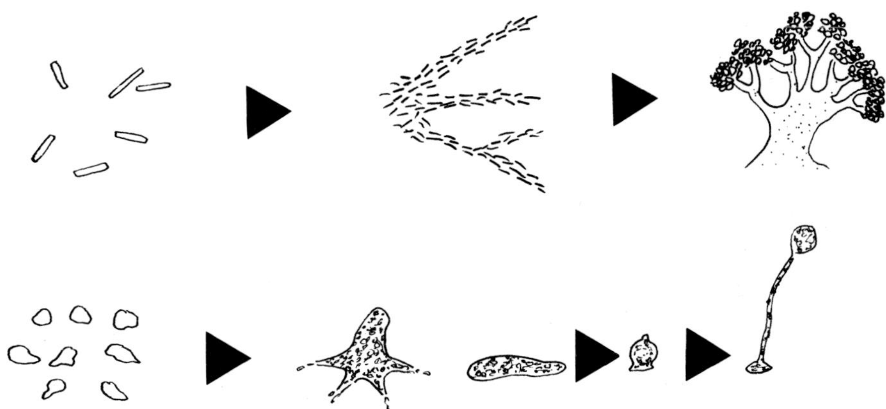

Figure 2.11. Organizational parallels in cell gatherings—slime bacteria (top row) and cellular slime moulds (bottom row). Initially separate cells (left) aggregate together, migrate (centre) and transform into a fruit body (right). (Not to scale)

Even in assemblies of identical cells, processes that lead to the differentiation of a heterogeneous (non-uniform) structure are inevitably initiated. Those cells on the margin of the aggregation will form a boundary layer that insulates all other cells in the system from the external environment. This boundary layer restricts the dispersion of individual cells and in so doing enables the aggregation to operate as a coherent organization.

Dramatic and extraordinarily parallel examples of such organizations are provided by the cellular slime moulds, in which the cells are eukaryotic, and the myxobacteria, which are prokaryotic (Fig. 2.11). In both cases, nutrient-limitation in the environment results in the association of initially widely dispersed cells to form a mobile "pack". The pack eventually becomes stationary and converts itself into a kind of spore-producing structure or fruit body (see also Chapter 4).

Another impressive example of parallelism between prokaryotes and eukaryotes is provided by the "actinomycetes" and true fungi (see Fig. 4.12). In both of these groups proliferation results in the formation of apically extending, protoplasm-filled tubes, "hyphae", which do not separate from another but instead branch to form coherent systems known as "mycelia". Proliferation within these systems does not lead to division into separate cells, at least not until spores are produced. However, fungal hyphae may become divided into "compartments" by the formation of incomplete partitions known as septa. Mycelia lacking these partitions are described as "coenocytic", in common with certain algae which also produce branching filaments. In many fungi, a further important property of mycelia involves the ability of hyphal branches to fuse or "anastomose" with one another, so converting initially radiately branching systems into true networks. As will be discussed further in Chapters 3 and 4, this ability may be important in the production of the large-scale spore-producing structures or fruit bodies—mushrooms, toadstools etc—which for many people are the most familiar parts of these organisms.

Cellular networks of a rather different kind from mycelia are produced by a group of marine organisms, the "Labyrinthulales". These organisms produce networks of "slime tracks", tubes with slightly elastic side walls, within which individual spindle-shaped cells move around, associating and dissociating rather like people exploring a maze.

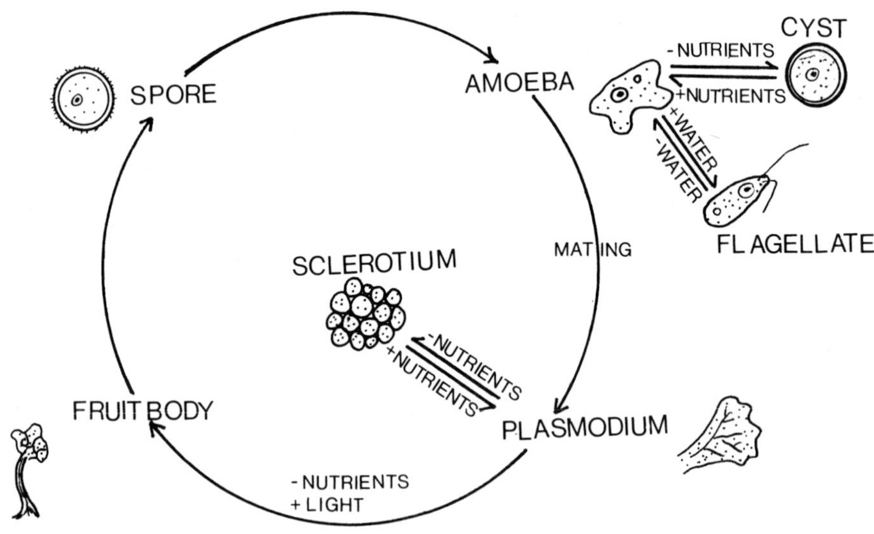

Figure 2.12. Life cycle of the slime mould, *Physarum polycephalum*. (Based on Carlile and Watkinson, 1994).

Mycelia expose the artificiality of rigid definitions of system units and boundaries. They are often referred to as colonies, but this undervalues the high degree of connectedness, differentiation and division of labour that occurs within them, not least when they form fruit bodies. On the other hand, to refer to them as multicellular individuals seems to be stretching a point because of the relative independence of their hyphae—which can be separated off and regenerated almost as readily as if they were free-living entities. Even to describe mycelia as multicellular is arguable since, although they contain many genomes, these can all occur within a single bounding cell membrane. In the true slime moulds this situation is taken to even greater extremes in the form of pulsating, multinucleate, slimy masses of protoplasm known as plasmodia that ooze around consuming bacteria or fungi before settling down to produce a fruit body (Figs 2.12, 2.13). All these systems might validly be described as extremely irregular single cells, but that would obscure their sub-divided as well as corporate organisation as societies of genomes. When it comes down to it, even the definition of a cell has its difficulties and can force thinking into inappropriate corners.

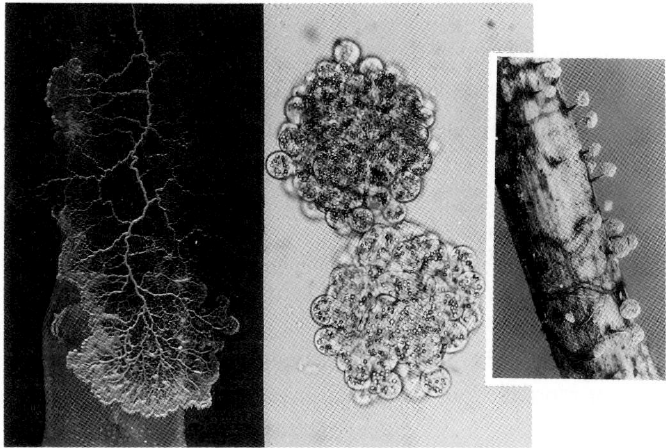

Figure 2.13. Organizational stages in the life cycle of slime moulds. Left, plasmodium of *Physarum polycephalum* migrating along and ingesting a smear of yeast cells. Centre, microsclerotia of *P. polycephalum*, consisting of numerous multinucleate spherules formed following nutrient exhaustion. Right, fruit bodies of *Physarum viride* on a dead stem—tracks left by plasmodia as they migrated to sites of fruit body emergence are visible. (Left and centre from Carlile, 1971; right courtesy of John and Irene Palmer).

When, then, does an assembly of cells (presuming these are clearly definable) qualify as a fully-fledged multicellular organism? It cannot just be when it begins to function differently from the sum of its parts because such emergent properties (see Chapter 1) are inevitable even in small associations due to the loss of freedom on at least some part of each cell's boundary.

Generally, if an association of cells is to be regarded as an entire organism, then there should be clear evidence for some kind of co-ordinating influence, causing it to behave as an integrated structure. The occurrence of division of labour and consequent interdependence between specialised cells, tissues or organs would also be expected.

However, there is simply no absolute decision point before which there is no co-ordination or interdependence and after which there is.

This point is illustrated by certain kinds of algae which develop either as unbranched filaments or as symmetrical (or nearly symmetrical) assemblages of cells (Fig. 2.14). An example of the former is *Spirogyra*, which consists of a single series of identical cells joined end-to-end. There is no specialization of function between the cells, and elongation of the filament involves elongation and division of cells anywhere along its length. The filaments can therefore be regarded as a linear series of cells which associate simply as additive "modules". *Oedogonium* is similar to *Spirogyra* but is more heterogeneous, with growth and reproduction occurring in distinctive regions or cells along the length of the filament and the basal cell being modified into an anchoring "holdfast".

Non-filamentous cell assemblages occur within algae known as the Volvocales. The cells within these assemblages resemble those of a free-living genus, *Chlamydomonas*. A sticky substance, mucilage, prevents the cells from dispersing, so that they become associated in composite structures known as "coenobia". The complexity of these coenobia varies. In *Gonium*, they consist of flat plates of sixteen identical cells. In *Eudorina* they consist of spheres containing sixteen to thirty-two identical cells. In *Volvox* itself, up to several thousand cells, all joined together into a network by cytoplasmic interconnections, encompass a hollow sphere, parts of which are specialized for reproduction, and which, as a whole, shows a high degree of co-ordination.

There are also parallels between the eukaryotic algal forms just mentioned and the prokaryotic blue-green bacteria (cyanobacteria), which can exist either as clusters of cells within gelatinous sheaths, or as unbranched chains of cells. The chains of cells show some division of labour because a single filament may contain cells specialized for spore production, for attachment to surfaces and for conversion of nitrogen gas into ammonia, as well as photosynthetic cells.

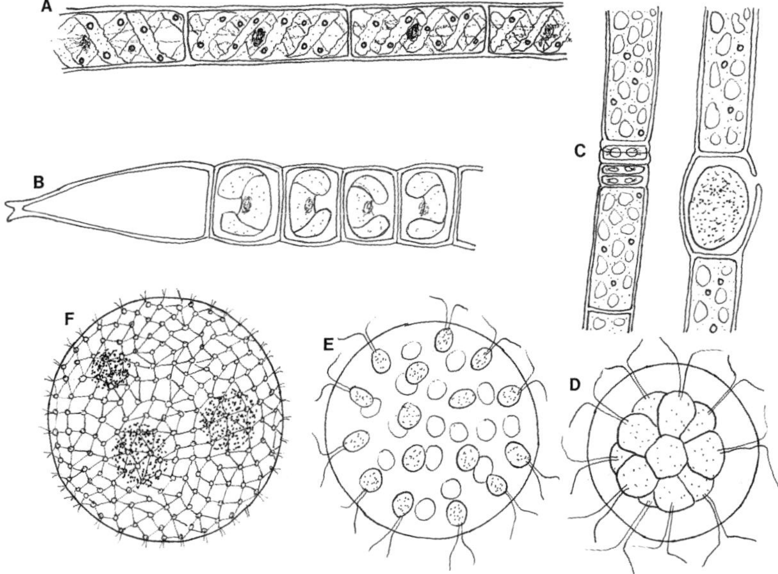

Figure 2.14. Thread-like and spherical cell assemblages of algae. A, part of a filament of *Spirogyra*, showing spiral chloroplast (stippled) and associated "pyrenoids" (storage centres), with nuclei suspended by cytoplasmic strands; B, part of a *Ulothrix* filament showing basal holdfast cell and open-ring-shaped chloroplasts; C, parts of *Oedogonium* filaments, showing vegetative cells with net-like chloroplasts and sperm- and egg-producing cells (left and right); D, coenobium of *Pandorina*, showing closely packed cells, each with two flagella, encased in a mucilaginous matrix; E, coenobium of *Eudorina*, with dispersed biflagellate cells; F, coenobium of *Volvox* with many biflagellate cells linked by fine protoplasmic connections—three daughter coenobia are present inside the parent colony. (Not to scale)

Another group of organisms, this time generally regarded as animals, in which the boundaries between colonialism and multicellularity are not easily defined, are the sponges (Fig. 2.15). The bodies of these organisms contain five basic cell types. These cell types are so arrayed as to produce structures, strengthened by "spicules" of calcium carbonate (lime) or a fibrous protein ("spongin), which are able to drive water currents through their simple to intricately infolded hollow interiors. These structures may in turn occur either as isolated units, or as irregular aggregations. A parallel situation occurs in the sea squirts or "tunicates", aggregations of which not only have common internal cavities, but even share vascular (blood) systems. However, whereas sea squirts are generally regarded as irreduceably multicellular organisms, it has long been known that if a sponge is broken up into its cellular components by pressing it through a woven material, these cells are able to re-associate into a functional sponge within a few weeks. Together with the fact that the specialized cell types are not organized into distinctive arrays, known as "tissues", this ability has usually been taken to indicate that sponges are colonies of cells rather than multicellular organisms. If so, aggregations of sponge units must be colonies of colonies of cells!

"True" multicellularity, according to the most popular view, is found in its simplest form amongst certain aquatic organisms. Animal examples include the Cnidaria—the jellyfish and their allies. Plant examples include many seaweeds. In these organisms, the integrational processes leading to the production of a co-ordinated system of specialized parts are thought of as having been brought to completion, though not ultimate sophistication.

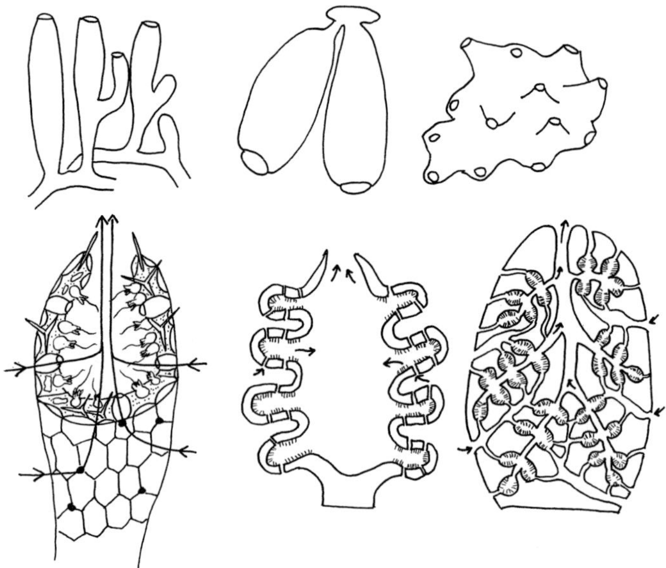

Figure 2.15. The structure of sponges. Top row (left to right): a simple ("asconoid") sponge, *Leucosolenia*; *Sycon*, a purse sponge; *Halichondria*, an irregular colony with many exit pores. Bottom row (left to right): part of a cut-away showing structure of an asconoid sponge with flagellate "collar cells" that set up water currents (arrows), pore cells, surface cells, amoeboid cells, spicules and gelatinous "mesenchyme" (stippled); longitudinal section showing body plan of a "syconoid" sponge with folded body wall; body plan of a complex ("leuconoid") sponge with elaborate internal channels. (Not to scale)

In the Cnidaria, cells specialized for contractile movements, sensitivity and the capture and digestion of prey are organized into two layers, an outer layer of "ectoderm" and an inner layer of "endoderm". These layers sandwich a middle, jelly-like layer ("mesogloea") within which a network of nerve cells serves to co-ordinate activities. The composite structure bounds a central sac-like, "gastrovascular" cavity, with a single opening or mouth, within which digestion and circulation of fluid takes place.

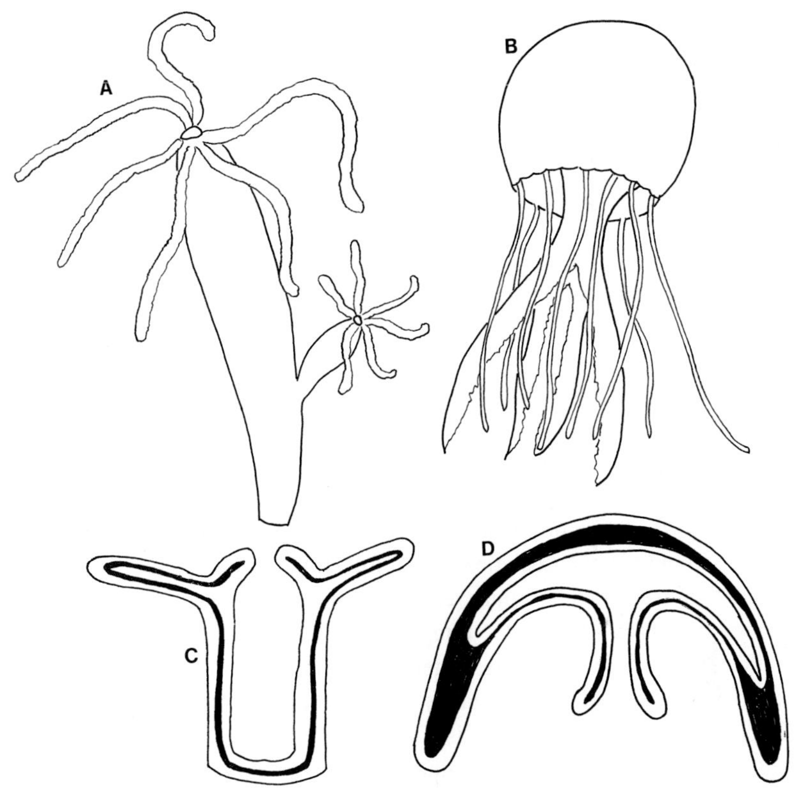

Figure 2.16. Polyps and medusae of Cnidaria. A, a polyp and a bud of *Hydra*, with mouth and tentacles at the top of a flask-like structure. B, a medusa of the jellyfish, *Pelagia*. C, body plan of a polyp in longitudinal section showing the jelly-like mesogloea (black) sandwiched between layers of ectoderm and endoderm. D, body plan of a medusa, showing how it can be interpreted as an inverted form of a polyp. (Not to scale)

The composite boundary of Cnidaria can be arranged into a variety of forms, of which there are two basic types—"polyps" and "medusae" (Fig. 2. 16). Polyps are stationary or have only limited powers of movement, whereas medusae are free to drift with the current and to shift themselves by a kind of jet-propulsion.

The Cnidaria possess tissues, in the form of interconnected arrays of muscle and nerve cells. However, as in multicellular plants, their specialized cells (other than those involved in reproduction) are not partitioned into distinctive local regions. This final level of multicellular sophistication involves the elaboration of the middle layer, or "mesoderm", between the ectoderm and endoderm, and the enfolding of tissues (see Chapter 4). It is found in those organisms, ranging from worms to dolphins, and including ourselves, that possess true *organs*. In such organisms, the requirement for highly organized neural, circulatory and binding systems (i.e. nerves, blood vessels and connective tissues) becomes paramount if behaviour and development are to be co-ordinated.

7. Societies of Multicellular Organisms

The suggestion made above that certain sponges could be regarded as colonies of colonies of cells may have rung one of those annoying repetitive alarm bells! It draws attention to the fact that no sooner have organized aggregations of cells arisen, than these aggregations themselves are prone to associate into higher order systems, culminating in the formation of complex societies. Also, echoing the progression to multicellularity, there are problems in making clear-cut distinctions between simple gatherings, colonies and societies of multicellular individuals. It is generally considered that true societies must exhibit significant division of labour between specialized activities that are integrated by means of various modes of communication. In extreme cases, where the components lose all their individual freedom, the society might appropriately be described as a "superorganism".

Simple gatherings, such as herds, flocks and shoals of multicellular organisms occur as inevitably, and for the same reasons, as do gatherings of cells. The organisms may be surrounded by some external barrier, they may be mutually attracted and/or they may fail to dissociate as they proliferate.

As with cells, the simplest gatherings of multicellular organisms consist of identical, or nearly identical "individuals". These individuals initially associate additively, i.e. as "modules", but as they do so, they produce a system which generates and becomes constrained by its own boundary and so is liable to develop as a coherent entity in its own right.

A good example of modular organisation is provided by certain sea mats (Bryozoa) known as "cheilostomes", which form crust-like coatings over rock surfaces, shells and seaweeds, and can sometimes even form frond-like structures resembling seaweed. The colonies of these organisms literally consist of aggregations of "little boxes", rather like the blocks of flats built in the twentieth century by human beings, in which resident individuals—"zooids"—live out their associated, yet compartmented individual lives. It is questionable whether the zooids really form a society; although there is some intercommunication between them through pores in the chitinous, calcium carbonate-impregnated walls of the boxes, these pores are usually plugged, so allowing only slow diffusion. However, some zooids have extraordinarily specialized forms: those known as "avicularia" because of their resemblance to a bird's head, may be specialized as "wardens", biting at intruders; those known as "vibraculae" serve as "cleaners", their long bristles being able to sweep away detritus or settling larvae (Fig. 2.17).

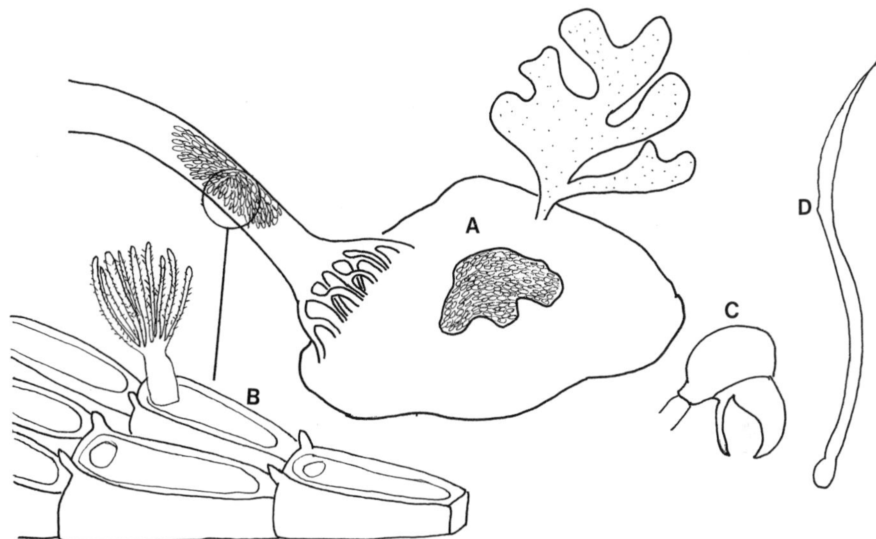

Figure 2.17. Modular organization of sea mats. A, a rock bearing a kelp stem colonized by *Membranipora*, a seaweed-like frond of *Flustra* and an encrustation of *Mucronella*; B, detail showing individual boxes, one with an emergent feeding zooid with ciliated tentacles; C, an avicularium; D, a vibracular bristle. (Not to scale)

Other Bryozoans inhabit more intercommunicating structures, and can hence be regarded as truly social. In "stoloniferous" forms, such as *Bowerbankia*, feeding zooids are attached to series of connective zooids arranged into tubes, in a manner very reminiscent of the interconnected compartments of fungal hyphae. In freshwater "phylactolaemates" there is free communication between the fluid-filled body cavities of the zooids, and in *Cristatella* the colony consists of a ribbon-like structure that is capable of slowly creeping over surfaces.

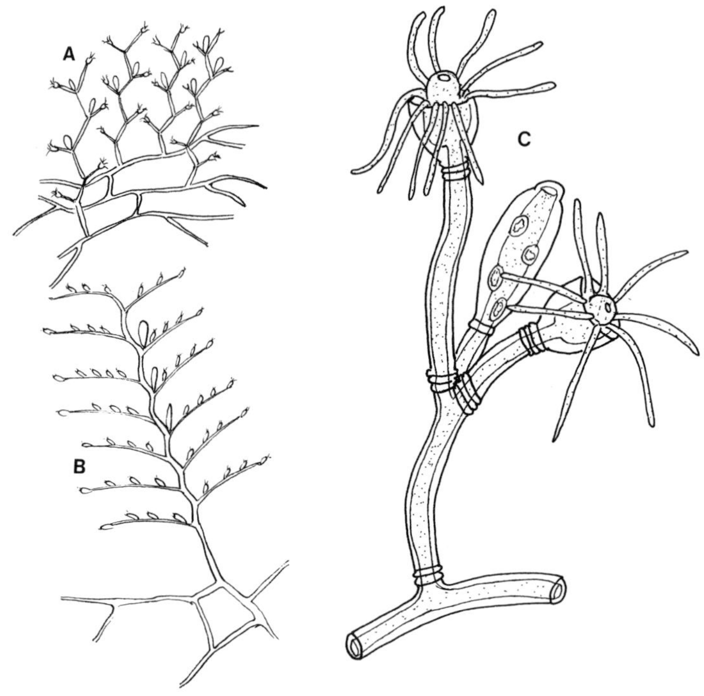

Figure 2.18. Social organization in hydroids. A, colony of *Obelia*; B, "frond" of *Plumularia*; C, detail of feeding (with tentacles) and reproductive polyps and perisarc of *Obelia*. (Not to scale)

Examples of truly social organisation are also found in certain Cnidaria—the branching "hydroids", the stony and octocorallian "corals" and the free-floating "siphonophores". Whereas the hydroids and corals are organized on somewhat similar lines to fungal mycelia, stoloniferous Bryozoa and higher plants, the siphonophores resemble animals with organs, the motile *Cristatella* (see above) or even a human fishing crew.

Hydroids consist of individual polyps whose central cavities are all connected to one another, usually by by a tubular system of erect, branching "stems" and creeping stolons or "hydrorhiza" (Fig. 2.18). The hydrorhiza extend outwards and give rise to further erect stems, so increasing the size of the colony until it reaches some physical boundary. Partially encasing the system is a chitin-containing layer, the "perisarc", which although composed of rigid material is capable of some movement because of the presence of corrugated articulation points at intervals along its length. This layer may have cylindrical, cup-shaped or vase-shaped openings, "hydrothecae", surrounding each polyp. An important feature is that whilst the polyps in any one colony are all genetically identical to one another, they occur in a variety of forms. Many are feeding polyps, equipped with tentacles that trap prey. Others are reproductive polyps, giving rise to medusae which drift away and engage in sexual reproduction. Yet others may be equipped with stinging cells that protect the colony or paralyse prey. *Velella* (Fig. 2.19) is a free-floating hydroid form or "chondrophore".

The siphonophores are complex assemblies not only of different kinds of polyp, but also different kinds of medusae, all interconnected to one another (Fig. 2.19). As well as having sexual functions, the medusae may be modified into "swimming bells", protective flaps, or a gas-filled float. A well-known siphonophore is the "Portuguese-man-of-war", *Physalia*, which bears a fishing crew as undiscerning between their fish prey and human swimmers as are some tuna fisherman between their quarry and dolphins! Another siphonophore, *Muggiaea*, tests boundary definitions to their limits. Depending on context, it may be viewed as a colony of colonies or a superorganism of superorganisms. Within the mantle of one collective are produced a series of smaller collectives ("cormidia"), each of which is capable of detachment and leading at least a temporarily independent life.

The stony corals are related to the solitary sea anemones. However, they characteristically secrete calcium carbonate from the basal portions of their polyps to form a stony platform or cup. Whereas some stony corals, such as *Fungia*, produce sizeable, solitary polyps, the majority are colonial, producing large numbers of polyps, a few millimetres in diameter, that are all interconnected by a sheet of tissue. This connecting sheet consists of upper and lower extensions of body wall bounding a continuation of the gastrovascular cavity.

Yet other Cnidaria, the octocorallian corals parallel the hydroids in the production of sometimes extravagantly branched structures like the sea fans and gorgonians (Fig. 2.20). However, unlike both hydroids and stony corals, they build an *internal* skeletal framework. This framework is covered by fleshy assemblies of polyps interconnected by tubular extensions of their endodermises that ramify through a common, jelly-like mesogloea, so forming a tissue known as "coenenchyme". In some octocorallians the polyps are all alike; in other cases they show functional specialization.

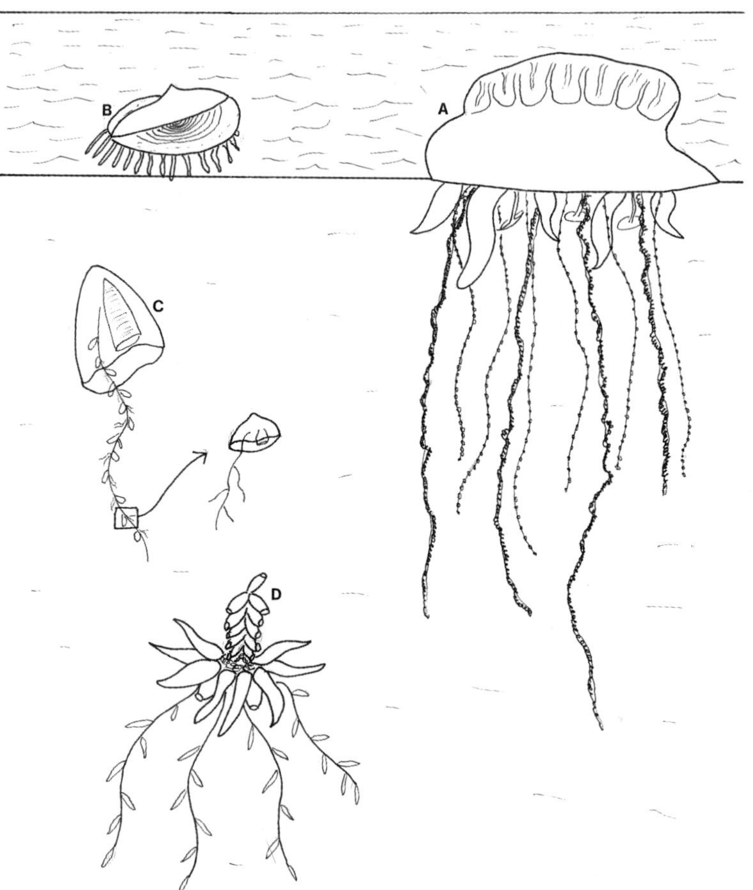

Figure 2.19. Social organization in siphonophores and chondrophores. A, *Physalia*, the "Portuguese Man-of-War" with its elaborate fishing tackle and gas-filled float. B, *Velella*, the "By-the-wind-sailor", with its sail and fringe of stinging zoids. C, *Muggiaea*, with its string of cormidia. D, *Physophora*, with its two rows of swimming bells, below which are feeding, defensive, fishing and reproductive polyps. (Not to scale)

Whether interconnected or not, the polyps and medusae of Cnidaria possess a body pattern that, like many flowers, exhibits "radial symmetry", i.e. more than one cut dividing the structure it into two equal halves can be made. However, many other animals, including human beings, are bilaterally symmetrical; they have distinguishable heads, tails and left and right sides.

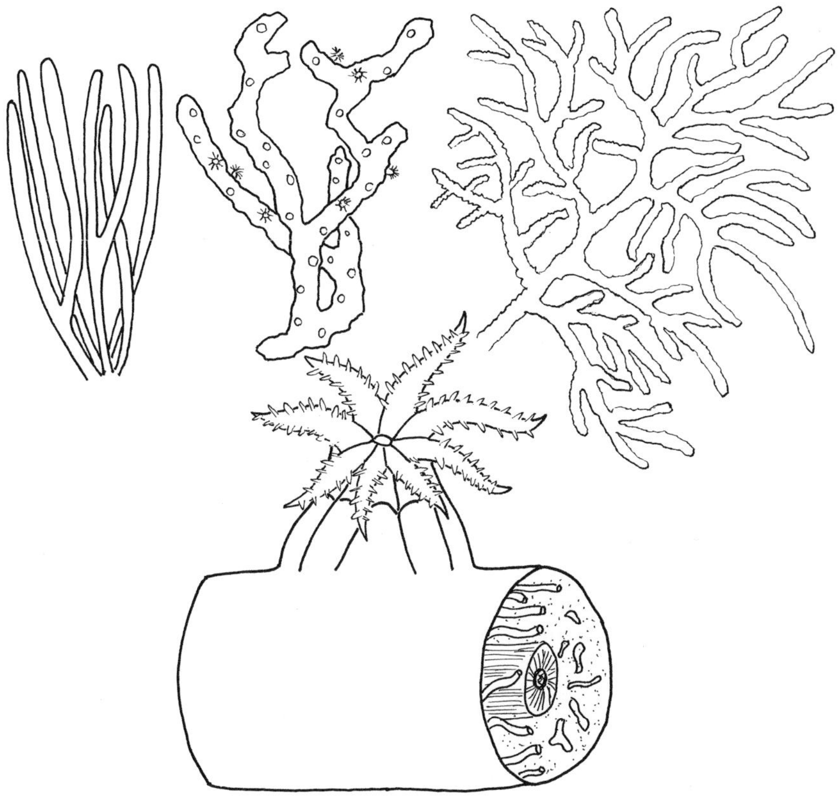

Figure 2.20. Social organization in gorgonians. Top row (left to right), sea whips, precious coral and sea fan. Bottom, detail of structure showing polyp with eight tentacles and main axis with internal skeletal rod and endodermal tubes ramifying through coenenchyme. (Not to scale)

Some of the simplest of these bilaterally symmetrical animals are the flatworms. It has long been known that these organisms, like sponges and hydroids, possess remarkable powers of regeneration in that they are able to establish new individuals if they are cut into pieces long enough for head and tail "poles" to be distinguished. One kind of flatworms, the tapeworms, are notorious parasites of human beings and other mammals. Tapeworms consist of an attachment structure, the "scolex", a narrow "neck" and a " strobila", consisting of a linear series of progressively larger body segments known as "proglottids". The worms are able to increase indefinitely in length through the proliferation of proglottids from the neck region. Each proglottid is a fully reproductive entity, and after fertilization may separate off from the strobila as a parcel of eggs. Tapeworms therefore have a similar kind of organization to filamentous organisms consisting of a series of near identical cells connected to a basal attachment point or holdfast: there is little specialisation along the length of the tapeworm. There might therefore be some justification for regarding a tapeworm as a colony, but most people prefer to view it as an individual.

A different kind of segmentation, known as "metamerism", is found in some other invertebrate animals, as well as in all vertebrates (animals with backbones)—including ourselves. Here, the repeated production of near identical body units during development is followed by the differentiation of at least some of these segments such that they can serve distinctive functions. In certain cases, as in many annelid worms, the number of segments is not fixed, so that an "individual" may continue to add further modules to its "tail" end, throughout its life. In other cases the number of segments is fixed, as in the adult forms of insects where differentiation into one head, three thorax (chest) and eleven (or fewer) abdominal regions occurs.

Insects are normally regarded as segmented individuals rather than societies of interconnected units. However, as segmented individuals they are certainly capable of organising themselves into complex societies paralleling, and perhaps even surpassing those of human beings in their coherence.

Insect societies clearly demonstrate the principle that a complex social structure is strongly correlated with a lack of independence amongst its members. This is evident in the fact that those insect individuals with the most sophisticated powers of perception and response to environmental stimuli tend to lead predominantly solitary lives—at least when they are out in the open. By contrast, members of arguably the most socially coherent insect collectives, those of army ants, are individually helpless—and blind or nearly blind.

Just as Cnidarian societies tend to be divisible into three basic components, with reproductive, feeding and protective functions, so insect societies can contain reproductive, worker and soldier castes. In Hymenoptera (bees, wasps and ants), the worker and soldier castes are female but either do not lay eggs, or lay eggs that remain unfertilized or are even used to feed growing larvae. They therefore only contribute to succeeding generations by tending the brood produced by the reproductive female(s), the queen(s), or by giving rise to males (see below). Males fulfil a role in mating with queens that disperse from the society following a "nuptial flight", but play little part in the life of the society other than as socially parasitic "drones". This may be related to the fact that whereas fertilized eggs develop into females, unfertilized eggs produce males. Recently, much attention has been focused on the genetic consequence of this situation: providing the queen mates only once, or mates only with her sons following the nuptial flight, sisters will on average be more related to one another than to their brothers or mother. There is, however, increasing genetic evidence that mating with different fathers does occur, resulting in an increase in genetic heterogeneity within the societies. In termites, where both males and females develop from fertilized eggs, males play a more active part in society life, and the reproductive male or "king" consorts the queen throughout his life rather than having only a brief "affair" following the nuptial flight.

Insect societies therefore generally originate from family groupings of related individuals. The same is broadly true of those social groupings formed by vertebrates in which there is the clearest evidence of functional specialisation and division of labour: prides of lions, troops of baboons, packs of hyaenas, societies of meerkats etc. The boundaries between these groupings are both maintained by external constraints and reinforced by territorial aggression. Such aggression is an expression of nonself (or non-us) rejection that appears in diverse forms at all levels of social organisation where disparate entities or groupings have the potential to integrate (see Chapter 7).

On the other hand, increased mobility of populations and suppression of rejection responses can allow scope for the evolution of cooperatives that are genetically and culturally diverse. This has been the recent trend in human societies, and it is not surprising that it should have been accompanied by considerable tension. A vital factor in overcoming this tension has been, and will be, the emergence and transmission of a kind of heritable information that is capable of transcending distinctions between genes and cultures—recorded knowledge. However, the *communication* of this knowledge, as applies to the sharing of any kind of resource, depends on the absence of competition and consequent integration, but not disappearance, of boundaries.

8. Partnerships and Communities

All the social and multicellular associations discussed up till now have involved the same species of organism, and, for the most part, genetically related or identical members of those species. However, integrative processes extend well beyond species boundaries, bringing what may be very disparate organisms into communication and culminating in the formation of mutualistic symbioses and community networks.

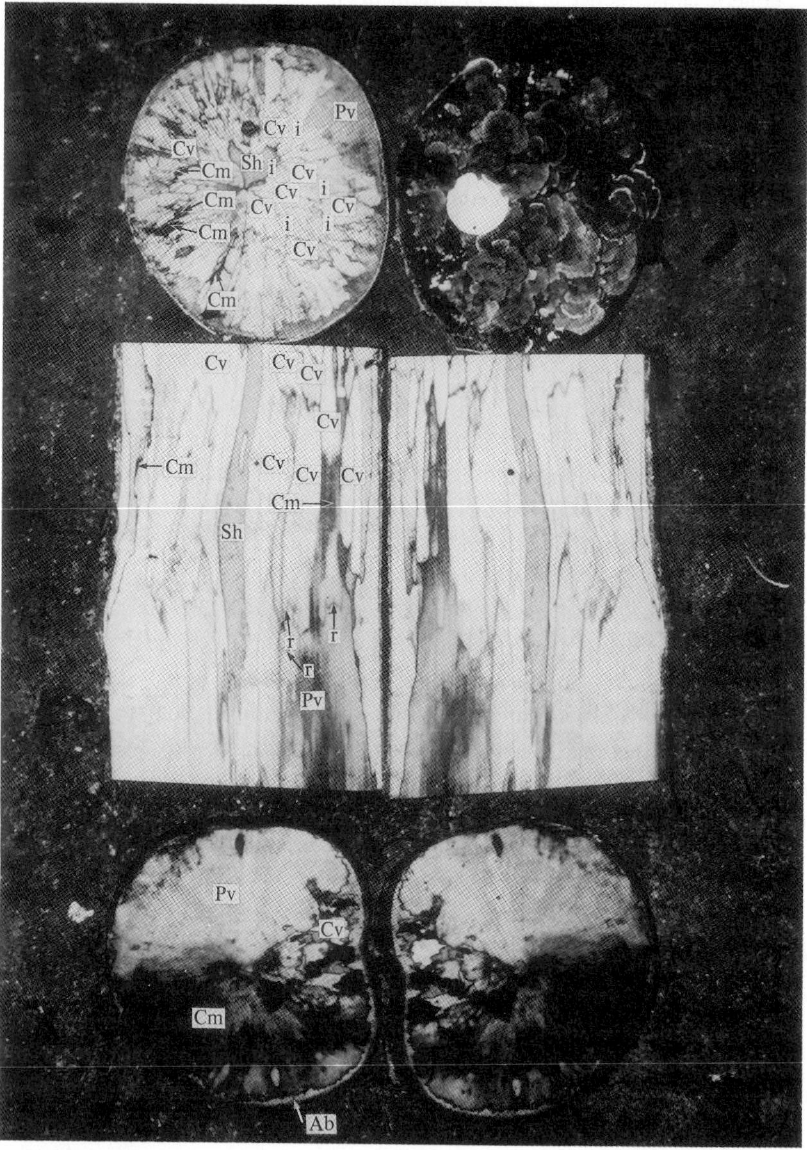

Figure 2.21. Embattled mycelia: fungal community organization in a decaying beech log which had been placed upright with its base buried in soil and litter two years previously in a deciduous woodland. The log has been sliced into cross sections near the top and bottom surfaces and into longitudinal sections for the

intervening length. The mosaic-like pattern directly reflects the spatial distribution of interactive boundaries (i) between competing and coexisting "individual" mycelia of the same and different species of fungi. The upper surface is covered by fruit bodies of the "many-zoned bracket" fungus (known in America as "turkey tails"), *Coriolus versicolor*. The distribution and action of some of the fungi present in the log is indicated by the following symbols: Ab, *Armillaria bulbosa* ("honey fungus") colonizing from the soil and occupying a peripheral region of "white rot" decay (due to degradation of lignin) just below the bark; Cm, *Chaetosphaeria myriocarpa*, a dark stain-causing, "poppies in Flanders" fungus that occupies the "war zones" between decay-causing individuals and other regions where decay fungi are not active; Cv, individuals of the white rot decay-causing fungus *Coriolus versicolor*; Pv, *Phanerochaete velutina*, a white rot decay-causing fungus which is invading from the base and replacing fungi that have colonized from the top, as indicated by "relic" boundaries (r); Sh, *Stereum hirsutum*, a white rot decay fungus. See also Figure 6.20. (From Coates and Rayner, 1985).

As mentioned in Chapter 1, far from being exceptional, mutualistic symbioses are major contributors to life on the planet. In terrestrial habitats the roots and root-like organs of the majority of higher plants (i.e. mosses, liverworts, ferns, cone-bearers and flowering plants) form intimate partnerships with fungi. These partnerships are known as mycorrhizas and are only absent from those plants which tend to colonize virgin or recently disturbed soils, such as those on rubbish dumps and at the leading edge of sand dunes. The fungi extend out, as mycelium, into soil, and thereby provide their plant partners with improved access to mineral nutrients and water in exchange for organic compounds produced by photosynthesis. The mycelium can also interconnect different plants—even of different species. By providing communication channels between the plants, mycorrhizal mycelia are thought to enable adult plants to "nurse" seedlings through fungal "umbilical cords", to reduce competition and to enhance efficient usage and distribution of soil nutrients. On the other hand, they can be piratized, as is demonstrated by the yellow bird's nest plant, *Monotropa hypopitys*, which by tapping into mycorrhizal networks is able indirectly to divert resources from the trees that participate in these networks.

Apart from forming mycorrhizas, the roots of some plants form associations with bacteria, blue-green bacteria and actinomycetes, that are capable of converting atmospheric nitrogen into ammonia by a process known as nitrogen fixation. These associations have considerable importance in the generation and maintenance of soil fertility, as anyone who cultivates leguminous crops (i.e. members of the pea family) knows.

Where higher plants are unable to establish themselves in terrestrial habitats, then surfaces that would otherwise be bare become covered by another kind of symbiotic union, in the form of lichens. Lichens are rather like sandwiches with the "bread" being composed of fungal tissue and the filling consisting of photosynthetic cells, either of green algae or blue-green bacteria. The union is remarkably tolerant of extremes of temperature and water availability, but very susceptible to industrial pollution of the atmosphere.

Not only terrestrial plants, but also many animals depend on intimate symbiotic associations for their existence. The guts of most animals contain assemblages of microorganisms and protists that both benefit from and can aid digestive processes. Some of these associations are indeed essential to digestion, and the activities of different members of assemblages complement one another. Such complementation occurs between the fungi and bacteria that inhabit the rumen of certain vegetarian mammals, and the bacteria and protists that inhabit the guts of lower termites. Some animals even cultivate partners that can aid digestion: amongst insects, these include the wood wasps, ambrosia beetles, higher termites and attine ants which not only use the fungi that they "domesticate" as a source of food, but also acquire enzymes from them that degrade such otherwise undigestible plant products as cellulose.

Symbioses are also of great importance in marine communities. The reef-building corals, for example, depend on the presence of photosynthetic protists, "zooxanthellae", within their tissues, and so cannot exist below depths where an adequate supply of light can penetrate. The corals benefit from this association through the provision of photosynthesized organic compounds, and the zooxanthellae are also thought to aid the production of calcium carbonate. The zooxanthellae benefit from nitrogen and phosphorus compounds in the food caught by the polyps, and may also gain some "shelter" within the animal tissues.

As discussed in Chapter 1, symbioses are relatively persistent, physical associations between organisms. However, even temporary associations are of great importance in the generation of interdependencies in natural communities. Complex "food webs" develop between photosynthetic plants as "primary producers" (i.e. the organisms which first convert the energy in sunlight into organic compounds) and carnivores as "secondary consumers". Arrays of organisms within these food webs provide sustenance for one another over a wide range of spatiotemporal scales. Although it might be thought that the activities of consumers would always be detrimental to producers, the consumers play vital roles in recycling and in preventing population growth from getting out of hand. Consumers also often play important roles in the dispersal of producers, acting as accidental or even deliberate propagators.

Despite the high degree of interdependency which can develop in natural communities, competition is also a predominant theme and has an immensely important influence on patterns of life, leading both to the establishment and breaching of boundaries. Some of the most obvious examples of the impact of competition on community organisation are found when non-motile organisms that potentially can expand and reconfigure their boundaries indefinitely encounter one another within the same living space (Fig. 2.21). A major theme in this book is that the patterns produced by such creatures vividly illustrate the dynamic processes underlying the evolution of contextual boundaries. These processes are obscured in those more discrete organisms whose short individual life spans necessitate the frequent production of further generations.

CHAPTER 3

DETERMINACY AND INDETERMINACY

1. Immortality and the Question of Growth or Reproduction

If they are to stay alive, all creatures—great and small—need continually to replenish the resources that they inevitably lose through their incompletely sealed boundaries. This need can be lessened if losses are reduced to a minimum by becoming dormant or restricting activity to levels just sufficient to keep the system "ticking over" within fixed boundaries—as in seeds, spores and hibernating animals. However, to thrive on a shifting environmental stage where (s)he who hesitates is lost, it is necessary to do more than just survive passively or maintain an "equilibrium state" in which energy gains exactly balance losses. No survival or equilibrium states can last forever because sooner or later they are always liable to suffer degradation, accidents, disease, competition or consumption in the uncertain world of dynamic boundaries.

Immortality, in the sense of the constancy imagined by Thomas Aquinas in *Summa Theologica* to apply to an angel, is therefore impossible. On the other hand, some form of *renewal* can always be assured as long as the ability of organisms to increase and change, that is to proliferate, is retained.

To theologians such as Aquinas, the ability of living things to proliferate has been viewed as a necessary evil in an imperfect world where absolute boundaries cannot be maintained. However, a different perspective comes from recognizing that absolute boundaries mean absolute stasis—a motionless universe in which time stands still! From this perspective, proliferation is vital to the ability of dynamic living systems to participate in an ever-changing scene.

The proliferation of living things is commonly described as one or other of two, seemingly distinctive activities—growth and reproduction. However, both "growth" and "reproduction" provide yet further examples of words like "individual" whose meaning is taken for granted and assumed to be precise when in fact they are open to variable interpretations. As might be expected, these interpretations depend fundamentally on where contextual boundaries are defined.

Where proliferation takes place within a contextual boundary, it is usually referred to as growth, but where it leads to separation of boundaries, and hence to the multiplication of "individuals", it is described as reproduction. This terminology therefore paradoxically equates multiplication (i.e. reproduction) with division! When, say, a single yeast cell enlarges due to the multiplication of materials within its interior, the process is referred to as growth. When it divides, following a phase of such growth, the cell is said to have reproduced. However, when division occurs within or upon a container or surface of some kind (recall the marble-filled bottle in Chapter 1), the process is regarded again as growth—but this time of the yeast population or colony. Likewise, cell division within a multicellular organism, if accompanied or followed by an increase in size of the organism's boundary, is regarded as growth. By contrast, processes where cell division is followed by isolation of the "daughters" from the parent, or where a multicellular structure becomes fragmented into parts that subsequently regenerate, are usually described as reproduction.

Whether proliferation is regarded as reproduction or growth therefore depends on where the system in question is thought to begin and end. As described in Chapter 1, this location depends on the standpoint of the observer—whether a system is viewed as an "object"—within an arena, or as a "subject"—the arena itself. Although this distinction may just seem like playing with words, it is in fact quite critical to understanding the ways in which life forms adjust to the uncertainties that arise from inhabiting changeable environments.

A living system that possesses an expandable or "open" contextual boundary that enables it to continue to *grow* and alter its pattern indefinitely, may be described as *indeterminate*. Barring catastrophes—such as a fire—big enough to encompass its entire boundary, this system is also *potentially immortal*. Even if it exhibits obvious cycles of degeneration and regeneration, the death of parts is essential to the process of renewal within the ever-changing boundary limits of the system as a whole. For example, the yellowing and detachment of old leaves is often regarded as a way in which resources are redistributed from redundant to actively growing parts of plants.

The link between indeterminacy and immortality may consciously or unconsciously underlie the powerful allure of concepts which seem to *interconnect* and so absorb the lives of mortal individual human beings into a self-perpetuating indeterminate system. These concepts include "family", "society", "civilization", "immortal souls" and the "collective unconscious". Individuals may come and go, but the system continues. Similarly, the human obsession with limitless economic growth may arise from a refusal to admit that there has to be an end to expansion because to do so would seem like denying the possibility of a future.

Why, then do we continue to contrive just about every conceivable means of bucking the system by asserting our individuality? Why too, do we latch on so readily the idea that individual competition, the antithesis of co-operation, is essential to progress?

This may be related to the fact that the central problem for us human beings as self-conscious individuals, at least when we view ourselves objectively and out of context, is that we epitomize *determinacy.* Determinacy is a condition common to many (but not all) animals, in which the body boundaries of "individuals" are localized: these individuals cannot be in two or more places at once and they inevitably die within a fairly fixed span of time. Such individuals may be free to move about, but they have little freedom to remodel their own body boundaries other than by moulting or adding on extra segments as they get older. Although animals which accomplish the latter, e.g. snakes and worms, are sometimes described as indeterminate, they generally repeat rather than remodel their previous boundary and so are better thought of as "modular" (see Chapter 2).

Since determinate individuals have a more or less fixed life span, they have to reproduce if their genes are to survive. This situation may, fundamentally, be related to the abilities of animals to *ingest* food and to locate further supplies using various means of locomotion. It contrasts with the indeterminacy of many plants, fungi, actinomycetes and colonial animals such as hydroids, which are not fully motile and which *receive* and/or *absorb* energy sources. These indeterminate organisms have at least the *potential* to grow indefinitely, until or unless they encounter some external limit to their expansion. Reproduction therefore provides them with the scope to disperse and reassort their genetic information (see Chapter 4) in the long term, rather than being an absolute necessity for individual furtherance in the short term.

In what are perceived to be discrete, mortal and therefore essentially determinate individuals, the sifting of successive generations by natural selection seems to provide the only means of adapting to changeable environments. The development of each individual seems to be predetermined by the genetic programme it inherits from its parent(s) and any ability it has to react to varied environmental conditions is embedded within this programme. The programme itself effectively uses the elimination of misfits in past generations to anticipate future circumstances. Significant alterations in its execution, or major departures from the future that this programme predicts, will not allow survival.

From this perspective, determinate individuals are defined by the genetic material that provides their only direct link with their parents. They are genetically and therefore developmentally "ineducable"—unable to change their genetic information content to correspond with novel circumstances—and so have "built-in obsolescence". They become viewed as self-centred objects that must react to their environments in such ways as to ensure the survival of their genes in their own or their relatives' offspring. Inheritable adaptations to new circumstances can only be achieved through a change in genetic information content brought about between the generations rather than by any kind of experiential process.

Remarkably, and perhaps reflecting our own individual determinacy, we human beings often expect such ideas to apply equally well to all living systems. Individuals are treated as finite units whose "adaptive fitness" can at least in theory be calculated and used to predict their contribution to future generations. Societies are thought to be organized most efficiently around central administrations. Evolution is interpreted as a calculational process that uses individuals, like beads on an abacus or bytes in a computer programme, to come up automatically with the best solutions to survival problems. Genes are cast in the role of potentially immortal "central controllers" and bodies as temporary contraptions assembled bit by bit (or byte by byte) according to a strict code of conduct.

However, the continuity and responsiveness of indeterminate systems defies such interpretations—and the fundamental significance of this fact becomes clear when it is realized that *at least some* indeterminacy occurs in *all* life forms. Just as there can be no absolute boundaries within or around any dynamic systems, so there can be no absolute distinction between determinate and indeterminate life forms. Rather, there are gradations and interactions between determinacy and indeterminacy, and so the terms can only be used relatively. Relatively determinate systems possess a modicum of indeterminacy and *vice versa*. Also, a system that may appear to be determinate from one standpoint can exhibit indeterminate properties when viewed in the appropriate context. For example, those many animals which have relatively determinate body boundaries follow and create pathways or "trajectories" in their surroundings as they change their position and behaviour over time and interact with one another. If these trajectories are mapped, they often exhibit a clear but irregular structure, similar to the body boundaries of indeterminate organisms like plants and fungi. This structure defines the dynamic context of the animals' lives, a fact which incidentally is of keen interest to animal trackers. Fundamentally indeterminate processes may also occur *within* the animals' body boundaries—in the development of veins, arteries and nervous systems, for example.

When it comes down to it, the indeterminacy, and hence the continuity, of any dynamic system requires both collective and individual action. Expandability depends on the system being bucked by individual action at its boundary—but if the bucking goes too far, the system will fragment. At the same time, the sustainability of the system in the face of temporal or spatial limitations in supply of resources necessitates integration of its parts to produce a less wasteful organization.

Ultimately, to treat any living part or parcel as fully determinate and discrete is therefore to divorce it from the dynamic context which it creates and interacts with, and hence to misunderstand it. The continuous mainstream of life flows through a context of indeterminate channels from which relatively discrete and specialized forms emerge as temporary, determinate offshoots, only to be resorbed when they die. Natural selection channels as well as sifts.

The aim of this chapter is therefore to uncover just what it is about indeterminacy that can so richly augment (but *not* replace) established ways of thinking about the organization and evolution of living systems.

2. Indeterminacy and Fluid Dynamics

So long as a system remains expandable, its boundary possesses some degree of freedom and so can continue to mould and be moulded by interaction with its external environment. The body boundaries of all organisms therefore exhibit at least some indeterminacy when young. This fact is used in certain human societies to "rectify" features thought to be undesirable—e.g. to straighten round shoulders, or to generate deformities such as extended lower lips or ear lobes that are deemed to be attractive or to confer status.

The deformability of indeterminate systems means that they can all be thought of as interactive "fields"—analogous to solidifying fluids—whose boundaries continue to expand along and create paths of least resistance in their environment as they gain and distribute energy. Given this common underlying theme, it is not surprising that striking parallels can be observed between the distributive patterns produced by indeterminate systems that differ fundamentally from one another in their scale and material composition.

The organizational patterns produced by an indeterminate, energy-distributing system arise quite naturally and spontaneously as the consequence of physical processes governing interactions between ingredients at and within its contextual boundary. Driven by energy input, but lacking a driver, the system automatically becomes organized into varied arrangements according to its local circumstances. True, the specific nature of these arrangements can be varied by altering the ingredients of the system. However, in that they *moderate* rather than *instigate* the arrangements, these ingredients—which in living systems are partly specified by genes—set the *parameters* or "limiting conditions" rather than provide the instructions for pattern-generation.

Fundamentally, indeterminate living systems therefore generate dynamic patterns primarily because of the way they are organized. Past activities directly influence future activities so that the system "learns as it goes along" through changes in its boundary properties, and the genetic code becomes more of a list of ingredient options and less of a chef. Genetic adaptation involves harnessing and fine-tuning the proliferative process, changing emphasis according to requirements, rather than changing the process itself. Development of the system can thereby be responsive to unpredictable patterns of change in the environment rather than being inflexibly pre-determined.

In order to grasp (if that is the right word!) the full significance of indeterminacy in the generation of life patterns, it is therefore necessary first to be aware of the fundamental organizational attributes of dynamic fields. Secondly, the varied guises in which indeterminacy manifests itself in living systems need to be recognised. Thirdly, it is important to comprehend how the patterns produced by dynamic fields are influenced by feedback processes. Fourthly, the mechanisms by which these feedback processes are brought into effect in individual cases need to be identified.

The first two of these issues will be addressed now, whereas consideration of the third and fourth will be delayed until chapters 4 and 5.

3. General Properties of Fluid-Dynamical Systems

The dynamic properties of a bounded fluid, and any system or dynamic field that can be likened to one, arise from the counteraction of expansive and resistive processes. Expansive processes result from the tendency of energetic entities to space themselves apart from one another following input to the system. Resistive processes can be due to various kinds of mutual attraction between system components, such as those which result in the surface tension of water droplets or the solidification of a cooling lava flow. Alternatively, there can be an actual physical envelope around the system. This envelope may either be "self-assembled" from materials carried within the system itself, or some kind of external barrier. Providing there is a means of supply, or input, the system will leak or deform at those points on its boundary where the resistance to penetration or expansion is lowest. In the process, flows will be established whose power and direction will both be determined by and be capable of reinforcing and weakening the boundary, so preserving and changing its configuration with the passage of time.

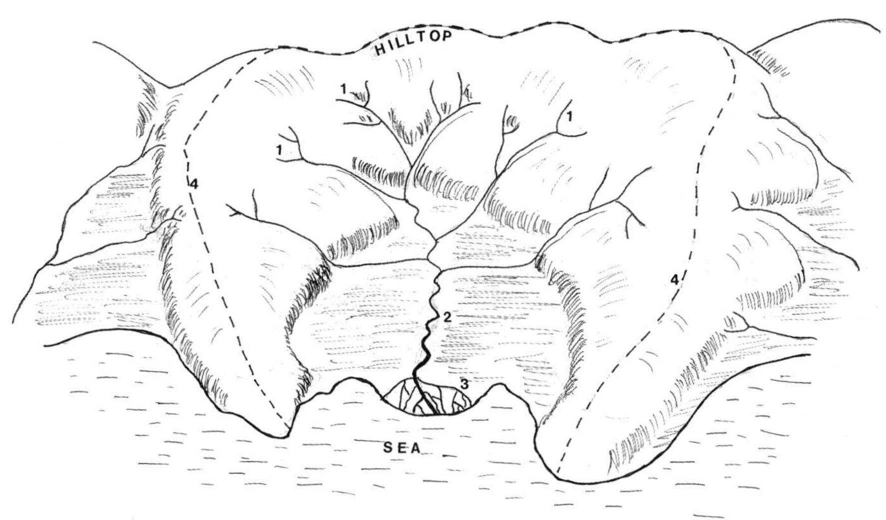

Figure 3.1. Characteristic features of a river basin. 1. Tributaries in headward-eroding valleys. 2. Meanders on the flood plain. 3. Distributaries forming a delta. 4. Watershed (dashed line) separating the system from adjacent drainage basins.

A familiar example of the kinds of pattern that can emerge from such processes is provided by a river system (Fig. 3.1). A general feature of these patterns is the presence of a connective channel between sites of uptake or *assimilation* where energy in various forms is gathered from the environment, and sites of *distribution* where it is relayed to another system or returned to the environment. The channel has resistant or "well-insulated" lateral boundaries that seal it off from the environment, so reducing both uptake and loss along its length and increasing its efficacy as a conduit.

In the case of a river, this pattern is due to precipitation of water (as rain, snow etc) onto a sloping landscape. Were the landscape to be completely uniform in slope, hardness and porosity, and the input of water continuous and too great simply too soak away, then an even sheet of water might flow across the landscape. However, such conditions are not met in reality, where there are inevitable discontinuities in input and irregularities in the landscape. Consequently, the run off of water from the landscape follows, reinforces and reconfigures some routes in "preference" to others. At the collecting or "assimilative" end, the system expands by the "headward erosion" of valley boundaries, due to the cutting action of water moving under the influence of gravity. At the distributive end, a delta may form, where the river breaks up into a series of branches. The latter are called "distributaries", as opposed to the branches at the head of the river which are known as "tributaries". If a sufficiently steep gradient is not maintained between sites of collection and discharge, then the river may "meander" rather than maintain a straight course.

There are several important stages in pattern-formation by indeterminate systems, and these will now each be considered in a little more detail. Water courses will be used as a familiar example with which to illustrate general principles, without dwelling, for the time being, on the specific mechanisms by which these same principles are brought into operation in living systems.

3.1 Developing Polarity

Imagine pouring water from a bucket into a pile of sand. If this is done continuously and carefully enough, a puddle will form that expands gradually outwards in all directions, i.e. *isotropically*. Sooner or later, however, as input continues, a threshold is reached. The constraints imposed by the surface tension of the water and the resistance of the sand to deformation and penetration at the puddle's boundary then become insufficient to contain further isotropic expansion. Instead the displacement of the water becomes focused at one or more "crisis points" where the boundary begins to give way. As this happens, all further expansion of the puddle becomes directed towards the crisis points and apically extending rivulets emerge, which cut and reinforce channels for themselves through the sand. Further addition of water to the system is now distributed along the channels and the remainder of the puddle's boundary ceases to enlarge—it may even shrink back a little. The system has "broken symmetry" or *polarized*, enlarging by elongating rather than by swelling.

3.2. Rippling and Branching

Into a free-flowing (non-turbulent) stream, immerse a boulder large enough to cause a significant change in depth, but not so large as to deflect the course of the stream, break its surface or cause it to break its banks. The input to the system from upstream will remain the same, but where it meets the resistance to its passage posed by the boulder, it will be "forced" to maintain throughput by locally deforming into a series of waves or ripples.

Now dam the stream, using a resistive but penetrable and moveable material such as sand or silt—simulating what happens when a river deposits its load of suspended particles into a shallow or tideless sea to form a delta. So obstructed, whilst still having to distribute the same incoming supply from upstream, the force of the stream becomes focused at sites where the barrier will be more prone to give way, so generating a series of narrow-angled branches. Meanwhile, at the head of the stream, any input of water greater than that which can be distributed by the system will also result in expansion of the boundary by the proliferation of headward-eroding, wide-angled branches.

3.3. Coalescence, Anastomosis and Networking

As adjacent fluid systems expand due to continuing input, there may come a time when their boundaries come into contact. Should the material separating them be removable, they may then coalesce, such that their contents become continuous. This may happen when the systems are still expanding isotropically—for example, two circular puddles may coalesce to form a single, dumbell-shaped pool. Alternatively it may allow branches to fuse or "anastomose", end-to-end or end-to-side—so forming "networks", or even side-to-side, producing "fasciations". Such events can have several very different, and very important effects on the distributive properties of the systems.

One important property of pooled systems is that they have combined capacities but proportionately reduced surface areas through which uptake occurs. They are therefore less prone to reach thresholds, where they break symmetry and branch, than isolated systems.

Such pooling effects depend, however, on the systems meeting on equal terms: in a river system this means that the beds of the tributaries or distributaries must be at the same level. If, on the other hand, there is a "potential difference" between the merged systems, then fluid will be redirected along the deeper or less resistant channel, so emptying the other channel. In river systems this results in the dramatic process known as "river capture"; small-scale versions of it can often be found in the drainage patterns formed by streams or an ebbing tide across a sandy beach and in the run off of rain drops from a window pane.

3.4. Scale Shifts and Degeneracy

When a branching system is converted, by anastomosis, into a network, the lowered overall resistance of the system will not only reduce branching but may also lead to a withdrawal of fluid from peripheral channels. These channels consequently become disused "cul-de-sacs" as the network develops into an increasingly powerful and self-sustaining *sink*. Energy then gets cycled around rather than distributed across the network. This may explain why the circular roads misguidedly built around metropolises in order to improve traffic flow can actually have the opposite effect.

Networks therefore have a strong tendency to "gridlock", to inhibit further expansion of a system's boundary. On the other hand, if a suitable outlet from them can be found, they are able to deliver far more power to the sites of emergence, allowing a fundamental change in the scale of expansion of the boundary. Correspondingly, a flooded river, by drawing on a large pool, bursts through a breach in a dam with far more force than would be possible if supplied through a single channel. Similarly, the larger than expected scale of some volcanic eruptions has recently been recognized to be due to enhanced supply of magma through underground networks formed by multiple channels extending along planes of weakness in the earth's crust. The same principle is well-known to electricians who recognise that far greater currents (electron flows) can be maintained, given the same potential difference (voltage), through circuits with resistances arranged "in parallel" rather than "in series".

However, even if suitable outlets allowing expansion from a network are available, the fact remains that continued expansion depends on maintaining supplies to the periphery, if the system is not to become self-limiting. As a network enlarges, the likelihood of self-limitation increases so long as the core of the system is maintained. However, if some mechanism, such as a one-way valve, can be put in place which isolates the core from further supplies, whilst still allowing redistribution from the core, then expansion of the system will become unlimited. The system will develop indefinitely as an outwardly expanding ring or "annulus" with an "explorative" margin and "degenerative" trailing edge encompassing a central "wasteland". There are many examples of such patterns amongst living systems, including some, such as city centre decay, which are the natural consequence of "economic growth" in human societies. A fungal example, the formation of "fairy rings" will be described shortly.

4. Indeterminacy in Varied Guises

Energy, and the boundaries across and within which it is transferred and distributed, can take a wide variety of forms. Sometimes it is contained in readily visible, material forms; at other times it can only be envisaged indirectly or even abstractly. Indeterminacy, due to the continued accumulation and distribution of energy within a deformable system boundary, can therefore appear in many guises, some more immediately obvious than others. A basic theme of this book is that whilst the specific mechanisms vary, dynamic processes of boundary expansion and restriction occur in all indeterminate systems. To provide some "feel" for how to think about these processes and the kinds of dynamically generated patterns that they can give rise to, I have described indeterminate systems as having "fluid-dynamical" properties. However, I don't want this metaphor to be taken too literally. The question I want to provoke is "in what sense can the pattern-generating processes in such and such a system be likened to those in a bounded fluid?". To begin to provide some answers, I will now introduce four superficially very different manifestations of indeterminacy.

4.1. Developmental Indeterminacy in Organisms (and Parts of Organisms) that Branch

Indeterminacy exhibited during the expansion of what are normally perceived to be the body boundaries of whole organisms may be thought of as "developmental indeterminacy". There may also be a case for applying this term to those tissues or cells that serve to interconnect different parts of an organism (e.g. blood and nervous systems) or to provide interfaces for exchange with the external environment (eg within lungs). It may even be apt to apply it to certain subcellular components such as membranes, mitochondria and cytoskeletal elements.

All organisms begin life as, or at least have an ancestry that can be traced back to, a single cell. This single cell can in many cases be referred to as a "zygote", the product of sexual fertilization. It is "totipotent" in that it contains all the genetic information that the organism will require during its life, and can give rise to all the various modes of expression of that information that occur within the organism as it develops.

As the cell proliferates, however, there may be a progressive narrowing down, associated with increasing specialization, in the capability of its descendants to express the full range of genetic information that they contain. The developmental fate of these descendants therefore becomes sealed by a process which in animals is known as determination and in other organisms is referred to as commitment. The more rapidly and irreversibly this process occurs, the more determinate the organism's pattern of development will be. By the same token, indeterminate development depends on being able to delay, localize or reverse the process, so that totipotency is retained indefinitely, at least in some part of the developing organism.

At one time the terms determinate and indeterminate development were applied to animal embryos, and thought to define two distinct groupings, the "proterostomes" and "deuterostomes". However, this distinction is by no means absolute, not least because determination is often a progressive rather than instantaneous process. Also, for present purposes, it is more useful to contrast the relative determinacy of organisms or parts that cease to grow, with the indeterminacy of those organisms and parts that maintain the ability to grow and change pattern throughout their life span.

A general property of relatively indeterminate systems is that they branch, having first become "polarized" by producing an elongating structure that typically extends at its tips. The branching structures themselves may be one cell wide, as in the hyphae of fungal and actinomycete mycelia, or they may consist of layers of cells or tissues, as in the blood systems of animals and the stems and roots of higher plants. In many cases the branches may also be capable of integrating, for example by means of hyphal fusions and root grafts, to form complete or partial networks. Alternatively, or additionally, they may grow in parallel, so forming cable-like structures such as the mycelial cords and rhizomorphs of fungi, the veins of plants and the nerves of animals. Degenerative processes may then ensue, allowing redistribution of resources along main channels to sites of continuing expansion; old branches end their lives feeding young ones.

These important processes in indeterminate development, namely polarization, branching, integration and degeneration correspond with patterns that are common to all kinds of fluid-dynamical systems (see above). They will each be considered in more detail in subsequent chapters. However, for a graphic illustration and foretaste of how they can be co-ordinated purely by local feedback, to produce energy-efficient, decentralized structures, I will now describe some fungal examples.

Fig. 3.2. Ordered pattern produced by the magpie fungus, *Coprinus picaceus* when grown in a matrix of 25 4 cm^2 chambers filled alternately with high and low nutrient media. Holes have been cut in the partitions just above the level of the medium. The fungus has been inoculated into the central, high nutrient-containing chamber, whence it has produced alternating assimilative and explorative states. Hyphal systems linking between chambers have been reinforced into persistent mycelial cords whereas others unable to extend further have been prone to degenerate. (Photograph reproduced by courtesy of Louise Owen and Erica Bower.)

Figures 3.2 and 3.3 show the kinds of pattern that can emerge when some kinds of fungal mycelia are grown amidst arrays of nutrient-rich and nutrient-poor sites. The patterns are initiated by the generation of profusely branched, "assimilative" (energy-gathering) hyphae in nutrient-rich sites. Non-assimilative, "explorative" hyphae are then driven across nutrient-poor sites. The connections between nutrient-rich sites are reinforced and resources are redistributed from degenerating hyphae through these connections.

Figure 3.3. Stages (left to right) in the development of a foraging mycelial system of the fungus *Steccherinum fimbriatum* between an already colonized beechwood block and an uncolonized beechwood block "bait" respectively placed in the centre and to one side of a tray of unsterile soil (from Dowson et al., 1988a).

Even where resources are less patchily distributed, the same basic processes of exploration, assimilation, networking and redistribution from degenerative hyphae can be found. In "fairy rings", the processes occur in an ordered sequence from the leading margin to the trailing edge of indefinitely expanding annuli (Fig. 3.4).

Indeterminate development therefore provides an effective means of expanding territory whilst continually adjusting to changes in local circumstances.

On the other hand, the production of structures most suited to such functions as photosynthesis, capture of prey, ingestion of food and reproduction depends on a high degree of adaptive refinement. The fact that many developmentally indeterminate structures give rise to determinate offshoots, e.g. flowers, fruits, leaves and fungal fruit bodies, can therefore be understood. The highly prescribed and sometimes elaborate form of these offshoots contrasts with the relative simplicity and flexibility of the axes from which they emanate. The offshoots and their readily identifiable functions commonly divert the attention of human biologists, leaving the indeterminate structures taken for granted. Yet it is the indeterminate component of the system that generates the offshoots. This component also regulates the interrelationships between offshoots by providing interconnective pathways whose variably deformable and penetrable boundaries outline the channels of communication within the system. Determinate superstructure is integrated by indeterminate infrastructure.

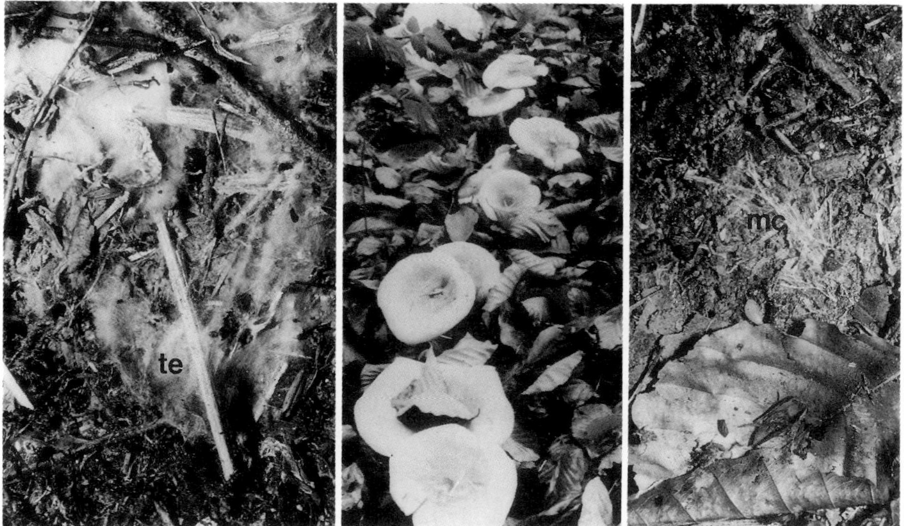

Figure 3.4. A line of toadstools (fruit bodies) produced by a "fairy ring" of the "clouded agaric" fungus, *Clitocybe nebularis*, consisting of an annular band of mycelium about 40 cm wide growing in beech leaf litter (centre photograph). At the advancing front of the band (right photograph), the mycelium is organized into exploratory parties of mycelial cords (mc) which is superseded in turn by diffuse nutrient-assimilating mycelium (left photograph) and a degenerating trailing edge (te). (From Dowson et al., 1989)

4.2. Social Indeterminacy in Nomads and Settlements

The fine distinction between developmental and social indeterminacy is epitomized by the "colonial" hydroids and bryozoa described in Chapter 2. The branching, anastomosing, hydrorhizal and stoloniferous systems of these organisms closely parallel the organizational patterns of higher plants and fungal mycelia. As in the latter, determinate offshoots with specialized functions occur, in the form of feeding, protective and reproductive polyps and zooids, interlinked by an indefinitely expanding tubular system.

In hydroids and bryozoa, the communication pathways are an organic part of societies whose individual members are sedentary. However, indeterminate social patterns can also be generated by motile organisms if they are mutually attracted and/or build or reinforce branching and/or labyrinthine pathways for one another to move along.

The pathways produced by motile organisms are prone to form very readily and without necessitating any kind of deliberate or pre-programmed calculation. The simplest case involves any assemblage of organisms in the act of migrating across terrain that in some respect can be regarded as "rough". The first organisms to enter the terrain, the "pioneers", make slow progress, but as they do so they inevitably smooth out irregularities, so creating paths of least resistance for others to follow. The followers, in their turn, continue to smooth and so reinforce the boundaries of the paths, so speeding progress along and reducing departures from "the beaten track".

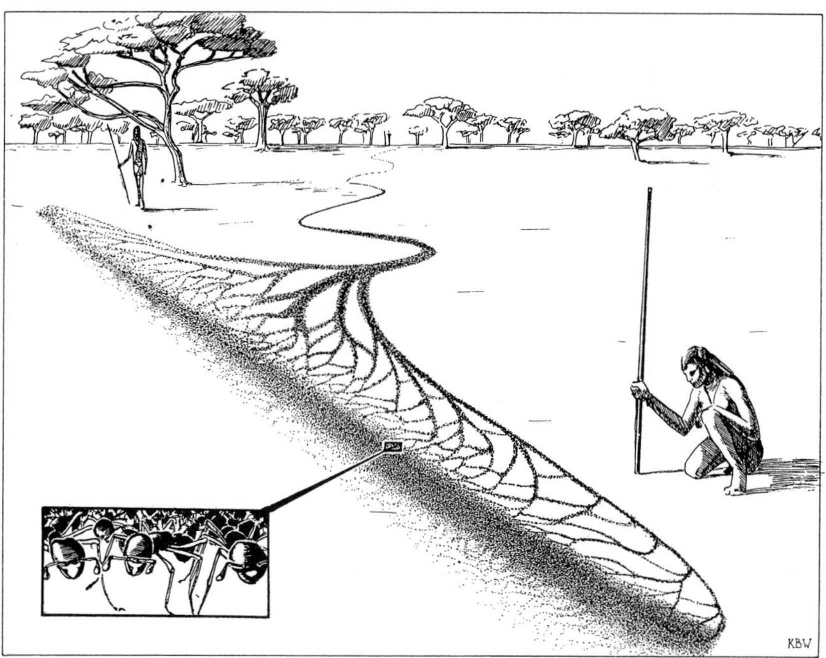

Figure 3.5. A foraging swarm of *Dorylus* driver ants produces a pattern like a river delta. (A drawing by Katherine Brown-Wing, taken from E.O Wilson, 1990)

As more and more followers speed along the trails, so they will eventually tend to catch up with and jostle the pioneers, often branching off to form new trails of their own. A structure resembling a river delta begins to form, as is graphically illustrated by migrating swarms of safari ants (Fig. 3.5) or herds of wildebeest (Fig. 3.6).

Whenever entities of any kind are impelled to cross some kind of resistive field, they will therefore behave as though they are attracted to one another due to becoming progressively constrained within the open-ended boundaries that they themselves create. This automatic pattern-generating process will, however, be greatly enhanced if the entities really are attracted to one another.

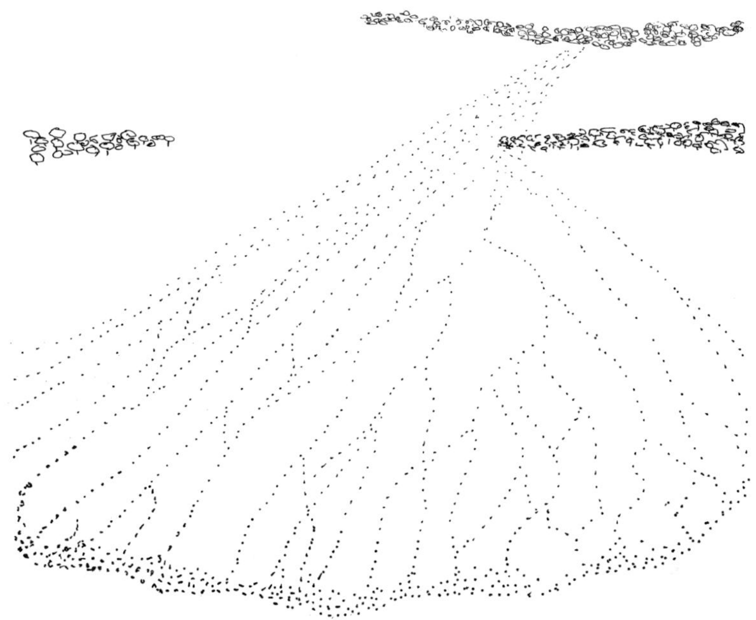

Figure 3.6. The "great trek"—a herd of wildebeest on the Serengeti plain in E. Africa migrates along well-worn trails towards river lands as the dry season advances. (Based on an aerial photograph).

An example of the role of actual attractants in pattern-generation occurs in the raid systems of army ants foraging for food. Here, individual ants deposit "trail pheromones", substances that attract other ants to follow where they have been. Over time, the tendency for the trails to be amplified by the positive feedback of successive ants depositing more on more is counteracted by the dissipation of the pheromones to the environment. Whereas usage of a trail, particularly when in both directions, therefore begets further usage, disuse begets disuse, causing it to degenerate as it loses its traffic. Foraging patterns emerge (Fig. 3.5) that are remarkably similar to those of fungal mycelia amidst patchily distributed resources (see above).

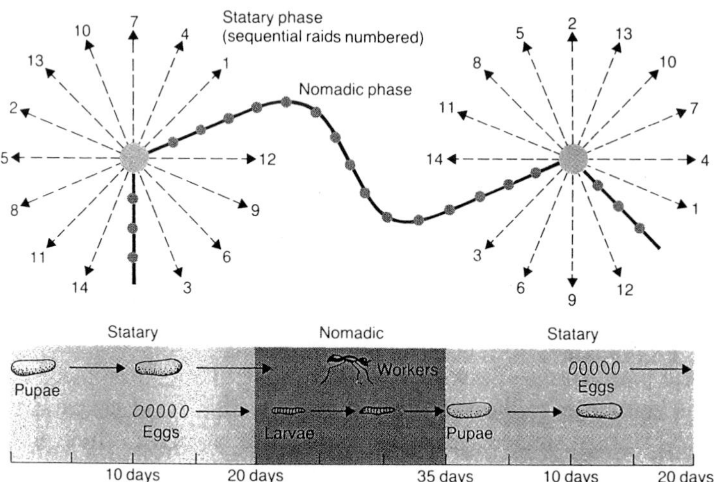

Figure 3.7. Symmetry and polarity in the 35 day foraging and migration cycle of the army ant *Eciton burchelli*. The spoke lines during the low-activity "statary" phase indicate the direction of raiding parties and the numbers the sequence of raids. Note that the angle between each raid is 126°. During the nomadic phase there is a new nest site almost every night. (From Franks, 1986)

Some army ants, such as *Eciton burchelli*, also vividly illustrate a pattern of life common to many hunter-gatherers, including certain human societies, which involves an oscillation between stationary and nomadic phases (Fig. 3.7). Whilst stationary, the colony is in an assimilative or energy-gathering mode, with foragers raiding a different sector of ground each day and returning to the same nest site each night. At this stage, the colony's brood consists of eggs and pupae, which don't consume food. However, when, with extraordinary timing, the pupae hatch into a voracious platoon of callow workers and the eggs hatch into larvae that eat as much per day as six adult workers, the drain on resources from the nest's already depleted neighbourhood becomes severe. The rate of return of food to the nest decreases and more and more workers join the foraging trail, recruited by the pheromone-laying of their predecessors. Eventually the entire population moves down the raid system to form a new nest at the foraging front. In this way the colony enters a nomadic or redistributive phase, taking it to a new feeding ground. The most intriguing thing about this process is the way the flow of outpouring workers develops its own momentum, preventing others from returning, and so turning the raid into a one-way street. Consequently the society is able to up-anchor, and move on.

Fully nomadic or free-roaming assemblages occur widely amongst living systems, ranging from packs of slime bacteria, cellular slime moulds and plasmodial slime moulds to herds of elephants. There may be relatively little differentiation within these assemblages, as in herds of mammals, flocks of birds and shoals of fish, or complex social structures with division of labour, such as those of Portuguese men-of-war or the bryozoan, *Cristatella*, may be formed, as described in Chapter 2.

The formation of undifferentiated groupings—herds, flocks and shoals—has attracted some attention from biologists anxious to find a suitably non-altruistic or "selfish" explanation for their formation. One interesting theory in migrating gatherings is that followers can cruise in the slipstream created by one or more leading individuals; the same principles are employed by human competitors in running, cycling and motor races. Two other theories focus on the dangers posed by predators; it is perhaps not surprising that these theories have been especially favoured, in view of the history of human beings as hunters. One of these, the "selfish herd" theory suggests, in effect, that individually selfish herbivores associate because of their need to "put some meat" between themselves and a predator. The other theory suggests that in groups there is an increased probability of at least one individual being alert to danger and so being able to raise the alarm. Consequently not all members need to be vigilant all the time and can get on with other things such as feeding.

Although these theories are attractive, and certainly do correspond with observable patterns of behaviour in groups, the advantages they describe cease to increase significantly (i.e. they become subject to the law of diminishing returns) for numbers above twenty. Yet many groupings are hundreds, if not thousands strong. What property of groups might be enhanced by such numbers?

Although this issue seems generally to have been neglected, it has sometimes been suggested that the complex dynamics of large groupings cause them to shimmer and fenestrate in such ways as to be intangible to a predator. This explanation may be getting closer to the truth. However, it still views the system from outside, as would a predator (or a human hunter), and accounts for the possible adaptive value of the patterns without actually saying how they arise in large groupings.

From preceding discussions it should be clear that combinations of large numbers of dynamic entities that are mutually exclusive and yet attracted to or constrained by one another will behave as dynamic fields. The larger the numbers, the more fluid-like will be the behaviour. The herd becomes like a pool of quicksilver set loose to roll about a landscape in which it circumvents obstacles, responds to intruders and detects paths of least resistance with extraordinary sensitivity. No wonder its members are elusive, and no wonder the primary "objective" of the predator is to separate off an individual whose movements can be followed more definitively. With very large herds, a requirement may even develop for groups of predators to co-operate with one another towards this end. As anyone who has watched a sheepdog in action will realize, predators, whether as individuals or as groups, are expert "irrigational engineers", being able to channel the movements of their quarry in appropriate directions. Crowds of human beings and traffic flows need be managed in the same way if they are not to seize up or dissolve into chaos.

What is it that binds the members of herds, flocks and shoals together, so that these fluid-dynamical properties are able to emerge? The roles of resistant external boundaries and direct attractants have already been mentioned. Another important consideration may be the self-insufficiency of the individuals. As has already been mentioned, autonomy of individuals runs counter to their tendency to associate. How, then, might self-insufficiency arise? Also, how might self-insufficiency be evolutionarily advantageous or at least be an inescapable consequence of the way living systems are organized?

An intriguing insight into these issues may be gained by thinking about herbivorous prey and carnivorous predators. The life styles of these organisms demand that they perceive their environment in distinctive ways, particularly with respect to vision. Herbivores, which are potentially under threat from all sides, require panoramic vision. Correspondingly they tend to have their eyes placed on the side of their heads, but at the expense of being unable to judge distance. Conversely, many predators, like ourselves, have binocular vision. This enables them to gain accurate, three-dimensional, positional information concerning the whereabouts of their prey, but makes them exceptionally vulnerable should they ever themselves become a target. Most of us know the discomfort that arises, when chased by real or imaginary pursuers, that comes from not having eyes in the back of our heads!

The lack of distance perception due to the need for panoramic vision in herbivores imposes unavoidable self-insufficiency. Rectification of this insufficiency by binocular vision would actually seriously disadvantage the organism. However, the insufficiency can also be overcome by associating with others whose behaviour and movements relative to one another provides reference points that enable the location of a perceived threat to be pinpointed. From this standpoint, it becomes clear that prey isolated from others by predators lose their reference points as well as their intangibility. Their only recourse then will be to behave randomly.

This example not only illustrates how coherent interactions can arise from deficiencies in individual perception, but also contrasts the highly targeted, energy-sapping operations of the predator, with the "systemic" responses of members of the herd. Being part of a herd involves some abandonment of self-determination and yielding to the influence of others. Such abandonment implies a loss of individual "responsibility", a diminution of alertness and an energy-saving readiness to "go with the flow". In human societies, the easing of conscience, if not consciousness, that comes from associating with others has great attraction and is common to all kinds of gatherings. Similarly, the relaxation that can be gained by meditating upon or absorbing oneself into any imagined or observed fluid dynamic movement—the sea, a snowstorm, a waterfall, clouds—can readily be understood.

Gatherings truly enable the blind to lead the blind. Individual decision-making is subsumed by group pressure, whilst the group is capable of organizing itself into an astounding array of interactive, energy-efficient patterns. Therefore, it is probably no coincidence that on the numerous separate occasions on which perhaps the most extraordinarily cohesive social systems—those of army ants—have evolved, the individual ants have severely reduced eyesight. Indeed some army ants possess only two single eye facets, contrasting with the elaborate compound eyes of most insects. The indeterminate society, rather than the determinate individuals that make up its numbers, becomes the vehicle for detecting and responding to environmental heterogeneity.

4.3. Phylogenetic Indeterminacy and Evolutionary Trees

Blindness is also a feature of evolution, the dynamic process of origins and extinctions that over many millions of years has generated the enormous diversity of living beings which have come to inhabit planet Earth. Some people view the evolutionary process to be predestined, and even go so far as to regard the emergence of human beings as its pinnacle of achievement. However, there seems to be little or no reason, other than conceit, to support such an idea. Indeed if evolution's "purpose" was to generate humankind, the path it has taken can only be described as incredibly indirect. It is always possible with hindsight to recognise how certain events have contributed to a current state of affairs, and thereby tempting to identify these events as the product of foresight or prescriptive fate. On the other hand, it is also possible to recognise how small changes in circumstances could have led, through feedack, to fundamentally differing long-term outcomes. When thought about in this way, the evolution of human beings appears to be just something that *has* happened amongst a myriad that *might have* happened.

In view of the predictably unpredictable course of evolution, and its sensitivity to small changes in circumstances, it seems appropriate that it has most commonly been depicted as an indeterminate form known as a "phylogenetic tree". In phylogenetic trees each ancestral line of organisms is represented as a stem which elongates through time. Any departure from these lines, recognisable as any set of organisms clearly originating genetically from and yet distinct from their ancestors is represented as a branch.

When a phylogenetic tree is examined at finer and finer scales, progressively more branches come into view. Trees intended to show only gross relationships between major groups of organisms are therefore represented as having relatively few, thick branches, whereas those intended to show finer detail trace a potentially infinite filigree. Extinctions are represented by branches which cease proliferating, and so cannot be traced to the boundary of the tree's "canopy" in current time. A measure of the relatedness of each grouping is indicated by the distance between their origins from the same ancestral line. Branches may diverge—become more different in character, or converge—become more alike. They can also grow in parallel, passing through similar sequences of stages during their evolution.

Phylogenetic trees therefore have much in common with the more familiar genealogies known as family trees. Indeed the latter can be thought of as the finest scale branches of phylogenetic trees. There is, however, one very interesting difference between the ways phylogenetic trees and family trees are usually depicted.

Whereas phylogenetic trees truly are "dendritic", i.e. tree-like, family trees are partial networks, due to mating between genetically divergent partners. Such networking, which involves the pooling of genetic information, is normally thought to be possible only within and not between species.

Nonetheless, it may be misleading to represent the evolution of species and larger groupings as a tree, firstly because of the occurrence of symbiosis and secondly because of increasing evidence for what has been called "horizontal gene transfer". As indicated in Chapter 2, symbiosis is a very widespread phenomenon and has had an enormous influence on the emergence of new niches, even if the partners retain a fully separate genetic identity. Horizontal gene transfer—the transfer of genetic information other than between sexual partners of the same species—has been demonstrated between bacteria, and even between bacteria and higher plants, by means of viruses and plasmids (see Chapters 2 and 4). It is also now occurs regularly in the laboratories of human beings indulging in genetic engineering. More speculatively, there would seem to be ample opportunity for transfer between symbiotic partners of all kinds, especially those engaged in the most intimate and interdependent associations. Indeed, there is ample evidence of extensive transfer between the genomes of mitochondrial and nuclear partners in eukaryotic cells (Chapters 2, 4).

Perhaps, then, evolutionary progress may prove to have been less tree-like than is usually imagined. Instead, it could be more like a foraging mycelium which extends, branches, anastomoses and degenerates as it gains, conserves, distributes and recycles energy in heterogeneous environmental settings.

4.4. Mental Indeterminacy in Learning and Problem-Solving

In many discussions of human and animal behaviour, a major issue has concerned how much an individual's actions and responses are innate, and how much they are acquired from experience; the so-called "nature-nurture" debate. A related issue concerns how much of an individual's behaviour is conscious and how much unconscious.

Here, there has been a strong tendency to equate unconscious behaviour with "hard-wired" nervous mechanisms. These mechanisms are fixed during the genetically pre-determined developmental programme that carries an embryo from conception to adulthood.

Conscious processes on the other hand can be thought of as more open-ended and therefore more susceptible to modification by interaction with a variable environment. They thereby facilitate learning and an ability to approach novel problems in a flexible way rather than by application of some prescribed, predictable routine.

Since consciousness is generally inferred to be a fundamental component of what it is to be human, it follows that human behaviour is very substantially influenced by the environment in which upbringing occurs. Indeed, it has been argued that human genes that "prescribe a large brain", and hence provide for plasticity and imagination, actually enhance the importance of nurture relative to nature in a developmentally determinate organism. Moreover, the indeterminacy of a human consciousness that is liberated from the rule of genes—for that is what these ideas imply—is envisaged to be the source of free will and individual responsibility.

The intuitive appeal of the idea that at least some human thought processes are indeterminate is evident in such widely used phrases as "streams of consciousness" and "lateral thinking". Most people, when asked how it feels to think about a complex, unfamiliar problem might well describe something akin to a foraging fungal mycelium. Thoughts radiate out from some inner space, as though searching out easier passages and circumventing obstacles, eventually cross-connecting with one another and becoming focused along particular channels. This kind of thinking has sometimes been described as "water logic", and often yields varied solutions that depend sensitively on circumstances.

Such indeterminate thought patterns contrast markedly with the kind of determinate, step-by-step calculation used when a problem is familiar and clearly defined. Such problems can be solved by applying prescriptive routines or formulae that have proved successful on previous examples of a similar kind. There is commonly a single, demonstrably correct solution that does not change.

There therefore seem to be two contrasting kinds problem-solving—the one prescriptive and precise but inflexible, the other innovative and versatile but prone to wander. Correspondingly, two very different kinds of approaches have been used to develop problem-solving skills: training and education. It is important to realize that these approaches are complementary, and reinforce different kinds of brain function. Which approach is most apt in particular cases will depend very strongly on circumstances. However, there has always been a tendency to regard one approach as superior and hence to give it more emphasis than the other. Generally, it is education that has lost out, perhaps because it is more indeterminate and harder to define, even though it is often viewed as "intellectually superior".

Training is appropriate to the performance of certain specialist tasks and actually requires a kind of blinkering to reduce the risk of error arising due to following indeterminate mental paths. Information is provided only to the extent which is vital to being able to perform the task, and the emphasis is on learning to apply a particular routine to exact specifications. The trainee is pre-programmed in the manner of a dedicated digital computer. The outcome is an obedient individual able to perform the task with speed and precision and without making mistakes. All is well until the unexpected happens or the task becomes changed or superseded in some way, in which cases the individual will be unable to adjust and may become redundant.

Education seeks to develop conceptual awareness and understanding. It does this by providing the individual with a range of background knowledge, together with experience of applications of that knowledge in a variety of situations. Mental indeterminacy is therefore preserved, if not enhanced, whilst providing some contextual boundaries along which thought patterns can be directed in a logical way. The educated individual is therefore self-disciplined rather than subservient, and so is versatile, quick to adapt to new tasks and able to solve novel problems. However, since no educated individuals can have complete knowledge or experience, thay are inevitably prone to error and idiosyncracy.

Trained and educated individuals are therefore self-insufficient for different, and fundamentally complementary reasons. Those who have been trained can apply specific techniques but may lack adaptability; those who have been educated are adaptable but may lack technical competence. Of course these descriptions only apply to extreme cases, and most individuals fall somewhere in between. Nonetheless, it should be clear that considerable benefits may accrue if the two approaches, and those learning from them, are allowed to work side-by-side, without ranking which is "better". Imagination, diligence, skill and enjoyment can then be combined, whilst avoiding the dangers of dullness and lack of realism that surface when training and education are kept separate.

Although learning—whether by education or training—might be regarded very much as a conscious process that frees the individual from genetic pre-programming, it actually represents another route towards the constrainment of behaviour. This is because learning processes are extremely re-iterative; a piece of information, an idea or action is repeated until it becomes committed to memory and automatic. Anyone who has learned to ride a bicycle or drive a car will be aware of the transition between the conscious effort required to do the multiplicity of things necessary to maintain momentum in early stages and the "just doing it automatically" that comes with experience.

Re-iterative learning processes have the effect of re-inforcing mental boundaries, making it progressively easier to retrieve information or repeat actions, but at the same time making it harder to depart from accustomed modes of thought or conduct. Thoughts and behaviour become stuck in a rut, and it becomes more difficult, as the saying goes, to teach an old dog new tricks. Re-education and re-training therefore often necessitate the dissolution of old boundaries. These processes are painful, if not downright humiliating, especially in societies obsessed with competition, status and notions of self-sufficiency. Yet they are as essential to continuance as the interplay between differentiation and integration, degeneration and regeneration is in all kinds of indeterminate systems.

CHAPTER 4

DIFFERENTIATION AND INTEGRATION

1. Divided and United States of Being: The Combine Harvester Principle

Any complex task can be subdivided into stages. For example, converting an agricultural crop into saleable produce involves gathering a harvest and then sorting, preparing and packaging it in a storable and/or distributable form.

At first, the entire task may be accomplished by a generalist who is able to complete each stage with sufficient competence to achieve an end result. However, it is difficult for generalists to do everything perfectly, and indeed the attributes needed to do one stage really well may be completely inappropriate for another stage. For example, the physical robustness needed to bring in a harvest may be incompatible with more delicate work needed in sorting or preparation.

Efficiency can therefore be increased by attending to each stage separately, using specialist skills and tools. However, whilst the separation of specialist tasks may reduce interference and enhance precision, it can also reduce continuity.

Any further increase in efficiency will then depend on developing some form of "communication system", such as a conveyor belt, that enables the component stages to be combined in an appropriate sequence. One specialism can then take on where the other leaves off. So, by first separating and refining individual tasks, and then interconnecting them, i.e. by first differentiating and then integrating, the generalist pioneer farmer may end up with a combine harvester!

If taken to extremes, the process of refinement and interconnection of specialisms is not without problems, however. The interdependence of finely honed components means that should one of them fail, or the nature of the task change in some unpredictable way, the whole system will grind to a halt. To avoid such catastrophes, it is necessary—as in real combine harvesters—to have various kinds of general purpose components and by-pass mechanisms that can keep the system going. Real combine harvesters are therefore the product of a delicate balance between precision (specialism) and flexibility (generalism). So too are all truly reliable and adaptable complex systems built by people.

Like harvesting and marketing a crop, staying alive is a complex task that requires a delicate balance to be maintained between differentiation and integration.

It involves a dynamic interplay between processes of gathering, conserving, distributing and redistributing energy. Whilst these processes are complementary they can also interfere, so it is vital for them to operate at appropriate places and times. They are therefore regulated either flexibly, by circumstance-driven feedback, or rigidly, by genetic prescription.

This chapter is concerned with two kinds of questions about differentiation and integration in living systems—*why* and *how*? Why do these processes occur? How, in terms of underlying mechanisms, do they work at different levels of biological organization?

At the outset, I want to recall that there are two basic kinds of approaches to explaining the existence of biological phenomena—adaptational and organizational. The predominant approach over the last century has been adaptational. This approach is based on working out *why* having particular kinds of attributes should enhance prospects of evolutionary furtherance. The fundamental issue therefore concerns *what* properties are needed in order to maximize chances of evolutionary success, leaving on one side *how* these properties can actually arise.

Underlying the adaptational approach is a tacit assumption that selective forces are so powerful that any required standard or evolutionary goal will be inevitably be attained. Indeed, for any currently "successful" organism, this standard will already have been achieved. Many theoretical biologists therefore work by predicting what the *best*, or *optimal* solutions to particular survival problems would be, and then examining how well real-life patterns correspond with these solutions. However, by so doing they risk concluding tautologically that things are as they are because if they weren't they wouldn't be here!

As several people have argued in recent years, the adaptational approach is also unrealistic in that it does not take into account constraints arising from the way life forms are organized. These constraints may, for example, necessitate costly "design features" that impede arrival at optimal solutions to problems. Therefore, real life patterns may reflect the best available solutions but not necessarily the best conceivable solutions: there may be a better way, but finding it may mean having to abandon the one that has served so far and entering a dangerous phase of redesign. There is, in other words, a tendency for life forms to get stuck in their existing organizational frameworks, unless some avenue opens up that enables them to shift sideways onto a new track.

Very similar kinds of constraints may account for the inertia that develops in commercial companies forced by market competition into continuing to do in the short term what has made them successful previously. Executives in such companies may well appreciate that long term failure to innovate brings increasing risks of diminishing returns on investment and ultimate liquidation. However, they also realize that any loss of competitiveness in the current market poses a more immediate threat.

It therefore takes daring to decide to *innovate* as opposed to *refine* in response to competition. It may also be foolhardy because it is so difficult to have more than very limited foresight. In the absence of foresight, innovations tend to clarify the target rather than *vice versa*. Once a target is clear, all subsequent steps towards it represent refinement, often in an increasingly competitive field.

Under the yoke of competition, adaptation therefore means approaching some already defined standard as closely as possible, given present constraints, in order to survive. There are compelling, "do-or-die" reasons for exhibiting particular structural or behavioural patterns.

From this standpoint emerges the familiar harsh picture of evolutionary change as the consequence of a relentless struggle for individual survival. A worrying feature of this picture is that it is liable to spawn the antisocial idea that all manner of adversarial, uncommunicative behaviour is justified by self-interest in the pursuit of "progress". Indeed it has already done so. The individual is required not just to be *competent*, which is one thing, but *competitive*, which is quite another. Life threatens to become at best an exhausting competitive treadmill, at worst a holocaust in which there can be no holds barred and all kinds of social fabric are torn.

There is, however, a complementary, more optimistic and less immediately competition-oriented approach to thinking about organizational attributes and evolutionary processes. Here, attributes are viewed more in terms of what they might *enable* than what they *constrain*.

Living systems are seen in this light to be inventive *in themselves*, by very nature of their internal drives. Freedom from competitive pressure, such as occurs under the conditions of "selection vacuum" or "evolutionary playtime" when new niches emerge (see Chapter 1), therefore actually allows more possibilities. Evolutionary success then results from serendipity—emerging in the right context—rather than by means of ousting opponents.

For those who might advocate more market competition to enhance consumer-choice (and *vice versa*) in human societies, it is therefore as well to appreciate that selection pressure ultimately narrows choice! In a competitive market, the consumer inevitably becomes faced either with a monopoly or a bewildering diversity of limited subchoices (see Fig. 1.2), as any bewildered supermarket shopper will know!

Adaptive necessity is therefore the strict governess, not the nursing mother of evolutionary invention! Inventiveness is implicit, and has to be nurtured by integrational processes before being exposed to the vicissitudes of the market. To deny this nurture in the short term is to stall evolution in the long term.

The same principle applies to scientific research. The frontiers of science are often referred to as "cutting edges". Here, chastened by their mutual competition, teams of researchers armed with limited foresight and advanced technology follow the seams of knowledge in precise, predictable stages. Such research is often referred to as "targetted" or "near-market".

Targetted research is easy to justify in the short term and readily attracts financial support. However, on its own it is generally extremely costly in terms of human effort and equipment, and at best leads to refinement of existing products and concepts rather than new insights and inventions. It requires more perspiration than inspiration.

By contrast, another kind of research, sometimes referred to as "blue skies" or "pre-competitive" research, involves working in what are more like "marginal fogs" than "cutting edges". This kind of research is based on a desire, born out of a mixture of curiosity and hope, to explore uncertain regions. In these regions, the research path is indeterminate and the nature and significance of any discoveries that may be made along it are fundamentally unpredictable. Such research is often considered to be "high risk", partly because it is often confused with speculation (long term prediction) and partly because there is always the possibility in the short term that it will lead nowhere. It is therefore usually deprived of resources. However, what may be perceived as the short-term risk of this research also implies a long-term promise—the *inevitability* that new and important discoveries and insights will arise sooner or later. Short-termism of all kinds, if allowed to monopolize resources, holds the future to ransom.

From what I have been saying, adaptation follows rather than drives invention. Adaptational arguments therefore explain why things are favoured or discarded by selection, but not how they arise. Nonetheless, these arguments have long held sway and, whatever else might be said, it is important to understand how a particular pattern of life, once it has arisen, may affect the ability to proliferate. I will therefore discuss adaptational explanations of differentiation and integration before considering the arguably more fundamental question of how these processes are internally driven by what I will describe as "organizational impulse".

2. Why Disunite?

2.1. Compelling Reasons: Adaptation

For reasons discussed in Chapter 3, all living systems must proliferate in some way if they are to thrive. What, though, is the best way to proliferate?

One way of proliferating involves expanding in all directions (isotropic expansion). Any system which continues to expand in this way retains its original shape and symmetry; i.e. it just gets larger. However, the surface area of its boundary relative to its volume is inevitably reduced as the distance between the boundary and the core of the system increases. This makes it harder and harder to keep the interior adequately supplied with resources brought in from outside.

These limitations on isotropic expansion are the compelling reasons usually given for why living systems in general and cells in particular cannot just get larger, and therefore have to divide. Further reasons lie in the fact that without division it is impossible to produce specialized components or to disperse to new locations.

Where the products of division do not dissociate, prospects open up for them to follow distinctive developmental paths—to differentiate—whilst still being able to be interconnected and so divide labour efficiently. These prospects are fulfilled most completely in developmentally determinate animal systems where an internal diversification process results in the partitioning of specialized local regions (organs, tissues, cells). Here, elaborate respiratory and digestive interfaces develop, interconnected by nervous and circulatory systems and bounded by an external skin which limits but cannot prevent dissipation to the environment. Differentiation and division of labour also occur, but to a lesser degree, in developmentally indeterminate systems. These systems divide externally by producing branches and so increasing their absorptive surface, rather than or in addition to dividing internally.

Where the products of division do dissociate, then they are usually viewed as the outcome of reproduction. As described in Chapter 3, reproduction enables dispersible genetic survival units to be produced in distinctive generations. It therefore avoids the risks to individual furtherance of having all one's eggs in the same basket. Instead, each basket takes its own chances in the face of judgement by natural selection, and those best suited to the circumstances of the time will survive and give rise to their own generations.

Such dissociation, when accompanied by the production and selection of varied individuals, results in differentiation at the population level of organization.

However, it can only give rise to societies that efficiently divide labour on the basis of their differences, if it is followed by some re-integrational process that opens communication channels. As will be discussed further in Chapters 7 and 8, such re-integration can bring great tension and instability as well as great opportunity.

2.2. Organizational Impulse: Destabilization

All systems in which tendencies for dispersion of gathered energy are counteracted by some kind of resistance behave as dynamic fields (see Chapter 3) that are inherently "excitable". As these systems gain energy, and use the energy that they have gained to gain more energy, they approach thresholds where the rate of uptake begins to exceed "throughput capacity". In other words, they begin to take in more than they can distribute uniformly to existing sites of deformation or discharge.

When input exceeds throughput capacity, the systems cease to exhibit easily calculated exponential (compound interest) or linear (simple interest) rates of expansion and instead start to subdivide and become increasingly irregular. Ultimately, the irregularities may become so great as to render the future behaviour of the systems at specific locations both highly unpredictable and extremely sensitive to small changes in the conditions affecting input/throughput rate. At this stage, the systems enter the realm of what has been described as "deterministic chaos".

There are many familiar examples of the development of irregularities as a consequence of driving systems beyond their throughput capacity. These range from the turbulence induced by an oar that is forced too strongly through water, to a "bunny-hopping" car whose driver is unable to balance clutch and accelerator. In fact, since any excitable system is prone to subdivide, whether by splitting, branching or enfolding, there is no need to invent subdivision as a means of adapting to the needs for reproduction, differentiation and effective energy transfer in living systems. True, rates and patterns of subdivision may be modified adaptively by changing boundary properties, but subdivision in itself comes automatically, impulsively rather than compulsorily.

The impulsive route to destabilization of excitable, dynamically bounded systems can be traced mathematically using what are known as "nonlinear" equations. The fundamental feature of all these equations is that the relationship between energy input and distribution changes from direct proportionality as the content of a system increases and resistances to expansion become more and more constraining. For example, in systems bounded by a restrictive physical envelope, the equations represent the relationship between energy gain and displacement of contents (throughput) to deformable parts of this envelope.

One of the best known nonlinear equations is the "logistic difference equation". This equation has found a variety of applications, most notably in simulating population growth. Usually the populations in question have been supposed to consist of discrete, and therefore easily countable organisms. However, the same basic principles can be applied to any bounded system containing increasing amounts of mutually repelling entities.

The equation relates the number of individuals (x) in a current generation of organisms to the number of individuals (x_{next}) in the next generation in terms of the net rate of increase per head of population (r). In the simplest applications, each generation is treated as discrete from the previous one, i.e. there is no overlap or carry over of individuals from one generation to the next.

Were there to be no boundary to the population and therefore no limit on its numbers, then the relationship between x and x_{next} would be directly proportional and so could simply be represented by the linear equation, Eq. (1).

$$x_{next} = rx \tag{1}$$

If this equation is *iterated*, that is, each value of x_{next} that it gives rise to is used as x to calculate a further value of x_{next}, then with each round of calculations, the values of x_{next} will rapidly escalate towards infinity.

This relationship obviously cannot apply to real populations where losses due to disease, competition and starvation would rapidly set in as numbers grow larger than can be supported by the availability of resources. In other words, as x gets larger in real populations it will become increasingly prone to constrain its own contribution to x_{next}, so that the relationship between generation sizes become non-proportional (i.e. non-linear).

This situation can be modelled by normalizing values of x and x_{next}, that is allowing them only to be some fraction of a theoretical maximum population size. The negative effect of x on x_{next} can then be represented by subtracting rx^2 from rx, as in Eq. (2).

$$x_{next} = rx - rx^2 \tag{2}$$

This is a form of the logistic difference equation. Here, the potential for increase in x, due to the *proliferative drive* resulting from resource acquisition represented by r, is countered by the *negative feedback* of rx^2. When this equation is iterated from some low initial value above 0, rx^2 increasingly constrains the increase in x_{next}.

Whilst rx^2 has virtually no effect at low values of x, it completely stops further expansion of the population by the time that x is equal to $1-1/r$.

The value of x at which there can be no net further expansion corresponds with what is sometimes referred to as the "equilibrium" or "carrying capacity" (equivalent to throughput capacity), K, of a population. It is a representation of the "balance of nature" where the forces promoting and retarding population growth cancel one another out, so that a dynamic equilibrium state is maintained. For values of r between 1 and 3, the equilibrium population size ranges from zero to 2/3, and iteration of the equation from low values will result in an initial increase in x. This increase either leads directly to attainment of the equilibrium value if $r<2$, or, if $r>2$, to a series of progressively smaller fluctuations (i.e. "damped oscillations") above and below the equilibrium value (Fig. 4.1).

For values of $r>3$, however, the population is driven by its own proliferative power over a threshold where it becomes unstable, unable to attain a single equilibrium state and instead subdividing or "bifurcating" into a series of alternative states. Here, as r is increased, x values come to oscillate around first two, then four, then eight, then sixteen...2n values in a so-called "period-doubling" cascade (the "period" which is doubled is the number of generations intervening between populations of equal size). At $r=3.57$, the population becomes fully chaotic and at $r=4$, ranges unpredictably over all sizes between 0 and 1 (Fig. 4.1). The sequence of values appearing in the chaotic region appears to be random, even though being derived deterministically (i.e. using specific values of x and r), and is extremely sensitive to small changes in the initial values of x prior to iteration.

The production of such complicated behaviour from so simple an equation has been a source of wonder to many and contains some important object lessons for those who might think that the only strong science is fully predictive science. Firstly, it demonstrates the impossibility of accurately predicting the long-term behaviour of chaotic systems: due to counteractive feedback, even the minutest change in initial conditions may result in radical departures in the behaviour of the system over time. This, rather than the incompetence of human weather forecasters, is the fundamental reason for the long-term unpredictability of weather patterns, for example. Chaotic systems are "predictably unpredictable". Whilst the general limits of chaotic behaviour can be outlined readily, what will happen at a particular location becomes increasingly uncertain as the future in question becomes more remote.

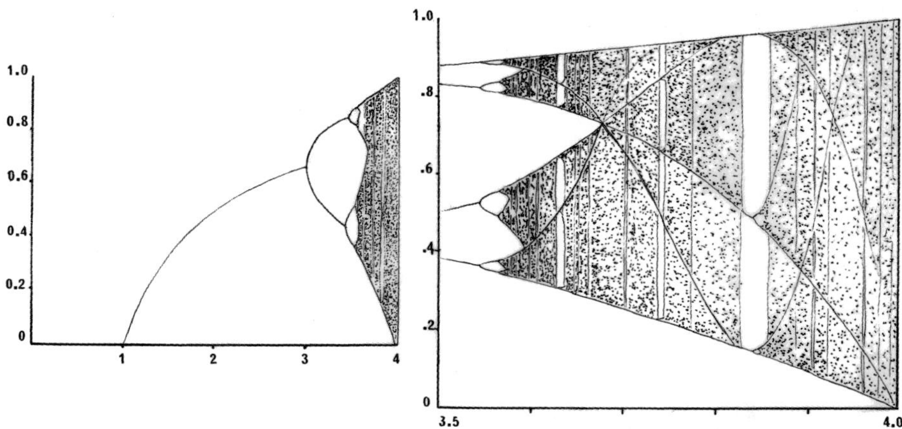

Figure 4.1. Population sizes (vertical axes) produced by iterating the logistic difference equation using different r values (horizontal axes). At r-values above 3, period doubling leads to chaotic dynamics above 3.57, interspersed with "windows of stability" (shown in more detail in the graph on the right).

A second point of interest is that systems that proliferate most freely, i.e. with high r values, are the ones most prone to instability—to "boom" or "bust". Many evolutionary theorists, however, equate proliferative power with "fitness" and so view any mechanism which constrains or reduces proliferation as "costly", regardless of context. One such mechanism is sexual reproduction, which, through the requirement of a mate, may at least halve the ability of an individual to contribute its own descendents to the next generation, relative to clonal reproduction (see below). However, in nonlinear contexts, what is perceived to be a "cost" may in fact be a "stabilizer", a means of preventing fluctuations between excess and insufficiency. Interestingly, sexual reproduction tends to occur in populations inhabiting stable niches, which are described as "K-selected". Asexual reproduction, by contrast, is often a feature of "r-selected" populations occupying temporary niches or in organisms where physical environmental factors limit growth at the edge of a population's geographical range. This and other factors influencing the potential costs and benefits of sexual reproduction will be discussed further below and in Chapter 7.

A third important point is that two variables that are strongly interdependent (what could be more interdependent than generation sizes?) may seem to be *uncorrelated*. In other words, if observed values for the variables are plotted as points on a graph, the points will be scattered rather than aligned with one another. A very large body of observational and experimental science is aimed at detecting possible causal relationships between variables (e.g. between smoking and lung cancer) by first demonstrating that they are correlated. A lack of correlation is usually taken to imply that there is no relationship between the variables—that they are independent. Yet, clearly, in a system exhibiting complicated, chaotic dynamics, dependent variables can appear to be uncorrelated.

A fourth point is that not *all* values of *r* above 3.57 result in chaotic dynamics. For example, at r values around 3.83, a process of "period halving" leads to the opening of a "window of stability", i.e. an interval of r in which three repeated population values or states occur. A sequence of bifurcations or period doublings then leads back to chaos as *r* is increased yet further. This sequence of bifurcations occurs in a similar geometric pattern, but at finer scale than that at $r>3$. As will be described shortly, the occurrence of similar patterns on different scales is a characteristic feature of systems exhibiting what is known as "fractal geometry".

Some understanding of how this complicated behaviour arises can be obtained by plotting the solutions generated by iterating nonlinear equations on a kind of map. Each solution is represented by a position on the map whose co-ordinates define the state of the system in what is called "phase space". By joining the positions produced by successive iterations, a kind of "fate path" or "trajectory" is derived which describes how the characteristics of the system change as energy is fed through it.

For example, a simple way of depicting iteration of the logistic equation can be achieved by drawing graphs showing the relationship between xnext and x for particular values of *r* (Fig. 4.2). The graphs take the form of parabolas which climb from zero to a maximum value, $\frac{1}{4}r$, before declining symmetrically back to zero. If a line is drawn at 45° to the x axis, this can be used to find values of x and xnext during successive rounds of iteration. It can then readily be seen how for *r* values between 1 and 3 (below 1, the population becomes extinct), a trajectory can be traced which "homes in" on the equilibrium point at which the 45° line and the parabola intersect. With $r > 3$, however, the trajectory cannot home in this way. Here, it either cycles around the equilibrium point, forming one or more closed loops (the number of loops indicating the period), or circumnavigates this point indefinitely without exact repetition.

The three kinds of behaviour just described correspond with what are known as three kinds of "attractors". An attractor is a state towards which the trajectory of a dynamic system evolves as if drawn by a magnet. The simplest attractors are "fixed point" attractors, equilibrium states which once arrived at cease to change. The carrying capacities of populations proliferating at r values between 1 and 3 are such fixed point attractors. Where a trajectory ends up oscillating repetitively between two or more states, the attractor is known as a "limit cycle". Where the trajectory is non-repetitive, but nonetheless never exceeds certain bounds, the attractor is known as a "strange", "chaotic" or "fractal" attractor.

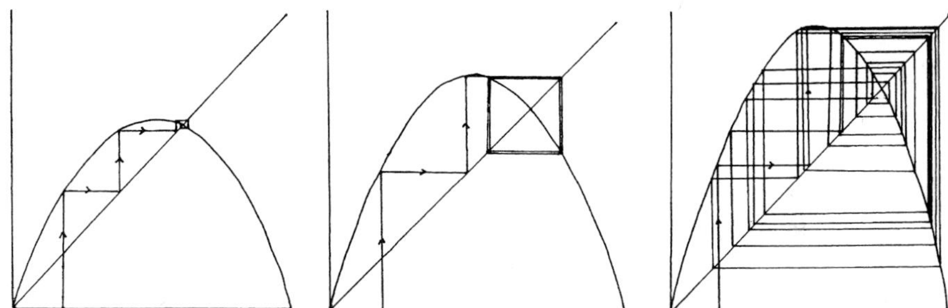

Figure 4.2. Trajectories produced by iterating the logistic difference equation using different r values. The horizontal axes represent values of x that serve as input, the vertical axes represent values of x_{next}. Arrows represent the process of iteration from initial values of x. Trajectories leading to equilibrium (fixed point, with r = 2.5), repetitive oscillation (limit cycle, with r = 3.1) and non-repetitive dynamics (chaos, with r = 3.8) are shown from left to right. (From Rayner, 1996b).

The concept of a "fractal" attractor, relates to a kind of geometry of irregular structures which cannot usefully be described in classical Euclidean terms of smooth curves and surfaces arrayed in integral dimensions of 1, 2 and 3. An important property of these structures is that when they are examined at closer and closer range, more and more irregularities come into view. Their lengths or areas are therefore infinite when viewed at infinitesimal scales, even though they can be circumscribed within finite planes or volumes.

What are known as "ideal" fractals exhibit exact "self-similarity", such that however small a portion of the structure may be, it will, when magnified, appear identical to larger portions. A classical abstract example of such a structure is the "Koch curve", which consists of an equilateral triangle, the mid-third of each side of which gives rise to a further equilateral triangle in an infinite regression (Fig. 4.3). Here, then, is a structure of infinite length contained within a plane. Individual parts of the structure are identical when viewed at different scales. For each "side" 3 units long, 4 units of material are present.

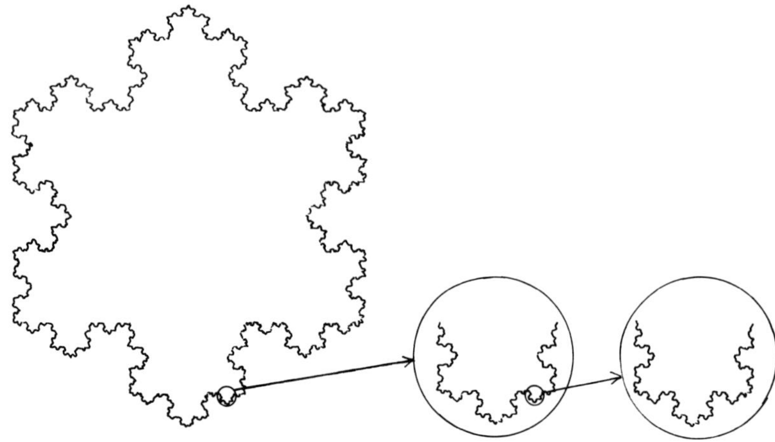

Figure 4.3. The Koch curve.

Though not conforming absolutely to this ideal, many naturally occurring structures are better approximated to fractals than to smooth figures. An example is the coastline of an island: any measurement of the length of the coastline tends to increase upon progressively closer examination as more detail comes into view. Similarly many kinds of branching structures are best approximated to fractals, at least down to a minimum scale corresponding with the width of their axes.

The problem of measuring fractal structures in a way that enables useful comparisons to be made is intractable if approached conventionally, using standard units of length, area or volume. However, it can be solved by measuring the degree of irregularity or "fractal dimension". Making this measurement involves accepting the idea that there can be fractional (hence, "fractal") as well as integral dimensions. This revolutionary idea has led to new ways of thinking about and defining all kinds of irregular or heterogeneous structures, from coastlines and boundaries around and between nation states to all manner of branching systems, animate and inanimate.

The basic idea of a fractal dimension can be grasped by thinking about the concept, already introduced, of a line of infinite length within a finite plane. Such a line is clearly more than just a line (with a dimension of 1) but less than a plane (with a dimension of 2). More formally, the amount of material, or "mass" (m) in a portion of the structure can be related to the radius of the field within which it is contained (r) by equation, Eq. (3).

$$m = kr^D \tag{3}$$

Here k is a constant and D is the dimension of the structure. For homogeneous structures, D is an integer and the density of the structure does not change with r. For example, for a plane, Eq. (3) becomes Eq. (4).

$$m = kr^2 \tag{4}$$

For a regular volume, Eq. (5) holds.

$$m = kr^3 \tag{5}$$

However, for heterogeneous, fractal structures, D is not an integer and density is dependent on r. D can readily be found for such structures from the relationship between r and m for different fields of view. In fact, if a graph is drawn of ln m against ln r, then the slope of the line gives D. The fractal dimension of the Koch curve, for example, is ln 4/ln 3 = 1.2618.

A very useful feature of the fractal dimension is that it indicates how uniformly a structure permeates space, by showing how the content of the structure (how much material it contains) is related to its extent (how widespread it is). For example, on a flat surface a branching structure with predominantly radiately aligned axes has a fractal dimension close to 1, corresponding with wide coverage and very uneven density, whereas a structure with D close to 2 has relatively more tangential axes and hence more even density. This kind of relationship is very important in biology because it has a considerable bearing on patterns of acquisition and distribution of resources in both competitive and co-operative organizations.

To summarize, fractal boundaries are prone to be produced around or within dynamic systems in which the rate of input exceeds that at which advancement and retardation processes can be maintained in equilibrium with one another. The systems therefore become divided into an increasingly complex series of subdomains as they are driven further from equilibrium. These subdomains provide scope for differentiation. However, the question then arises as to whether there is anything that can stop the process of subdivision from continuing indefinitely, on finer and finer scales?

3. Why Unite?

3.1. Complementary Attraction: Synergism and Co-ordination

The compelling reasons usually given for unification within or between system boundaries are synergism and co-ordination. Synergism involves bringing together complementary activities that work better together than on their own. Co-ordination involves timing activities so as to maximize their combined effect and reduce interference between them.

Synergism and co-ordination are therefore properties of systems whose whole is more effective in some respect than the sum of their parts. It is easy to see how these properties could provide an adaptive explanation for unification.

However, two more primary consequences of unification are the removal of competition and the minimization of dissipation of resources that occurs through over-exposed boundaries. Up until such time as communication channels are opened between neighbours, the most that can be expected from their proximity is a kind of neutral co-existence or non-interference based on the maintenance of well-defined boundaries between them. In practice, however, it is difficult to conceive how any energy-gathering (assimilative) systems in close proximity would not interfere in some way with one another's activities. In particular, they are prone to compete, the more so, the more similar they are to one another in their requirements for resources (see Chapter 1, Fig. 1.1).

In the absence of communication channels, competition and dissipation may either continue to the mutual detriment of perfectly matched opponents or result in the extinction of the less acquisitive entities. Competitive assemblages are therefore liable to be highly dissipative and unstable.

On the other hand, the unification of boundaries will allow an equitable distribution of resources between entities, according to overall supply and demand, so that less acquisitive components become supplied by more acquisitive components. This fact underlies the popular idea that, given the chance, less acquisitive individuals in human societies will become socially parasitic "spongers" off their more acquisitive ("wealth-creating") neighbours. However, when viewed in context it is obvious that less acquisitive components can never persistently take more than can be supplied by the system as a whole. Furthermore, whilst they may check growth in the short term, they can also play critical roles in exploring for, conserving and recycling resources (see Chapter 6). They therefore enable sustainable, persistent organizations to form in the face of temporal or spatial shortages in external supplies. If they then become isolated, allowing acquisitive components to gain monopoly, the system will feed on itself, leading to the disintegration of the whole or part of its structure.

The *disorders* that result from the proliferation of acquisitive components through the isolation and degeneration of non-acquisitive parts of a system are commonly regarded as being due to a loss of co-ordination. However, in the present context the re-establishment of competitive internal boundaries appears to be more primary. As will be described in Chapter 5, these disorders need not be a bad thing in indeterminate systems, where they enable redistribution of resources from redundant to active regions. However, they inevitably lead to the collapse of determinate systems, where they are commonly referred to as cancer or autodegeneration.

Just as unification of boundaries can have dramatic consequences, so too can the *timing* of unification have a critical bearing on these consequences. The sooner unification occurs following subdivision, the less specialized or differentiated can be the components, and so the less effective can be any division of labour between them. On the other hand, the longer that communication is delayed, or the sooner it becomes prevented between adjacent entities, the greater is the tendency for competitive disequilibria to develop. Similarly, the more specialized the subdivisions, the more self-insufficient, inflexible and interdependent they become. The balance between differentiation and integration is clearly very delicate, and where it is most appropriately set depends critically on circumstances. It can be varied in indeterminate systems, but any departure from it within a determinate system is liable to be catastrophic.

3.2. Connecting Attractors: Re-iteration and Breaking Moulds

As I have suggested, any system with a proliferative power sufficient to overshoot its throughput capacity is prone to subdivide indefinitely, following the trajectory of a strange attractor as inputs and outputs are iterated. By the same token, the propensity for the trajectory of a strange attractor to wander indefinitely depends on it never intersecting with itself.

On the other hand, any means of convergence that leads to the fusion, or anastomosis, of trajectory components will allow the exact *re-iteration* of pathways, as in limit cycles, and thereby prevent further proliferation. In the case of the logistic equation, such convergence can actually be brought about by an *increase* in drive, as during the period halving at r values around 3.83 (see above). The answer to the question, what is to stop endless rounds of subdivision therefore lies in the establishment of connections between the subdomains or branches.

When trying to relate the trajectories of attractors in phase space to the behaviour of real, physically bounded systems, it is useful to think about the effects of fusion on resistance to throughput. Without fusions, the resistances to throughput posed by a system's boundary are all in series, coming one after another. However, fusions convert the system, either wholly or partially, into a true network in which resistances occur in parallel.

Where resistances are connected in series, then the overall resistance to throughput is the sum of each individual resistance, as in Eq. (6).

$$R = R1 + R2 + \ldots Rn \tag{6}$$

Where resistances are in parallel however, it can be shown that the reciprocal of the overall resistance is equal to the sum of the reciprocals of each individual resistance, as in Eq. (7).

$$1/R = 1/R1 + 1/R2 + \ldots 1/R3 \tag{7}$$

The overall resistance of a system made up of a set of individual resistances (R1 to Rn) is consequently liable to be far greater if the resistances are connected in series than if they are connected in parallel. Networked systems can therefore maintain very much higher throughput rates before becoming unstable than serial systems.

Networking therefore represents a means of increasing throughput capacity and so restoring symmetry to a system that has started to subdivide, so that it once again becomes "self-contained". Persistent, self-reinforcing, seemingly "remembered" pathways are produced along which energy is distributed rather than dissipated to the environment. However, the pattern of distribution through a network is not as predictable as in a limit cycle. The trajectory of a limit cycle will always eventually return to the same point and then repeat itself exactly. In a network, there are many parallel options, and use of particular pathways may ultimately lead to disuse of others.

Although networks have resilient, potentially highly stable structures, they can nonetheless become unstable at very high rates of input, particularly so if some of their pathways are prone to degenerate (see Chapter 5). Once this has happened, very high rates of throughput can be maintained via a network to any "point of departure" where it subdivides or "breaks symmetry". Systems which break symmetry in this way therefore amplify the scale of their operations; individual conduits either enlarge themselves or become bundled into cable-like aggregates. The resulting synergism allows the boundaries of the system as a whole to become capable of expansion and redistribution.

Integration within and between dynamic systems therefore provides a basis for great stability, but at the same time allows previously imposed limits to be exceeded. In this way, fundamentally new, mould-breaking structures can emerge. These new structures are not possible without integration.

4. Mechanisms of Differentiation and Integration

Having discussed *why* differentiation and integration occur so universally, it is now time to consider the diverse *mechanisms* that cause these fundamental *processes* to operate in different organizational contexts. Here I want again to emphasize the point that whilst these mechanisms are very different from one another materially, they all achieve fundamentally the same ends in terms of association and dissociation. This point echoes the main theme of Chapter 2, and so the following descriptions will trace a path through organizational hierarchies similar to the one I took then.

4.1. Segregating, Transferring and Pooling Genes

The ways in which genes are encoded and copied within lengths or closed loops of DNA molecules were described in Chapter 2. My aim here is to portray how genes can disperse and reassociate within and between individuals and populations. Before I do that, however, it is necessary to describe how genes can "stay within bounds", by proliferating only within individual boundaries or "lines of descent" that arise directly from one another.

4.1.1. Maintaining Genetic Identity: Clonal Proliferation and Cell Division Cycles

An essential point to bear in mind when thinking about genetic distribution patterns is that genes can only multiply within cell boundaries (see Chapter 2). Equally, cell boundaries can only be maintained as long as they continue to harbour the genetic information needed for assembly of their components.

Where each cell generation is identical to the last, the processes of genetic and cellular proliferation are therefore tightly coupled in what are known as "cell division cycles" (or just as "cell cycles"). Given such tight coupling, an interesting issue concerns the extent to which it is the genes or the cell boundaries that are in control of events. Any effect of the boundary on input of resources to the cell will influence the rate of multiplication and division of the genetic material. Since the properties of the boundary are in turn determined by both genetic and environmental influences, there is likely to be a complex interplay between these influences. Following on previous discussions, determinate systems, may generally be more prescriptive and less subject to environmental influences than indeterminate ones.

Depending on where contextual boundaries are defined, cell division cycles result in growth or reproduction (see Chapter 3). Either way, the genetic information which is proliferated is essentially identical: in the absence of error, each cell generation receives the same complement of genetic information. The resultant sets of genetically identical cells or organisms can therefore be described as "clones".

Since any entity which is capable of multiplying has, by definition, proved its evolutionary fitness, clonal proliferation makes sense according to the short-term principle of never changing a winning team! However, this also means that the only way in which any genetic variation can arise is through errors—i.e. "mutations".

In bacteria, clonal proliferation is achieved by replication of the single circular chromosome and distribution of the progeny molecules, attached to the plasma membrane, into "daughter" cells that divide off from one another. There appears to be no special device necessary to ensure that each daughter cell receives an identical genetic complement.

In eukaryotic cells, however, where sets of linear chromosomes are bounded within a nuclear membrane, a more elaborate cell cycle occurs which has two distinct phases, known as "interphase" and "mitosis". During "interphase" the chromosomes are stretched out and not readily discernible as distinct entities using normal methods of microscopic observation. The DNA in each chromosome is replicated during a synthesis or "S" stage, which is preceded and followed by two apparently inactive stages or "gaps", "G1" and "G2". Following G2, the phase known as "mitosis" occurs, during which the duplicated chromosomes become thicker and shorter, and so more readily discernible, prior to the separation of each set of copies into daughter nuclei.

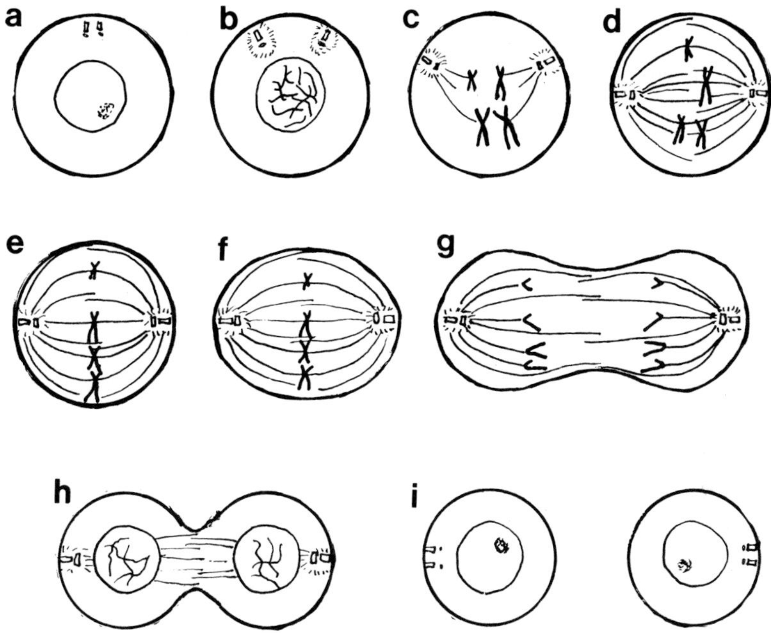

Figure 4.4. Mitosis. (a) "Interphase"; nuclear membrane and nucleolus present, "centrioles" shown at top of cell. (b) Early "prophase", chromosomes appear, centrioles begin to move apart and become surrounded by microtubules, forming "asters". (c) Mid prophase, nuclear membrane disintegrated, chromosomes thicken, spindle starts to form. (d) Late prophase, chromosomes associated with fully formed spindle. (e) "Metaphase", chromosomes arrayed at equator of spindle forming the "metaphase plate". (f) Early "anaphase", cell and spindle elongate. (g) Late anaphase, chromatids separate and move apart. (h) "Telophase", nuclear membrane reforms, cells begin to separate. (i) Interphase, two daughter cells with identical DNA content to the original cell.

Typical stages of mitosis are illustrated in Fig. 4.4. At the beginning of the process the duplicated DNA molecules (and associated histone proteins) are contained in sister "chromatids", joined to one another at a constriction known as the "centromere". The centromere region attaches, usually by means of a disc-like "kinetochore", to what becomes the "equator" of a framework of microtubules, known as the "mitotic spindle". Once the full set of chromosomes is arrayed at the equator, the chromatids draw (or are drawn) apart from one another by a mechanism which is still not fully understood, and thereby migrate to the poles of the spindle. Nuclear membranes then close around the now separated chromatids (=chromosomes), usually accompanied by the division of the original cell into two identical daughters. A new division cycle may then be initiated in each of the daughters.

Mitosis does not always result in the generation of identical daughter cells, however. For example, when nuclear proliferation is not accompanied by the delimitation of cell boundaries, multinucleate cells or "coenocytes" are formed. Also, where multiplication of chromosomes is not accompanied by delimitation of daughter nuclei, multiple chromosome sets occur in the same nucleus. Nuclei which contain multiple chromosome sets are known as "polyploid".

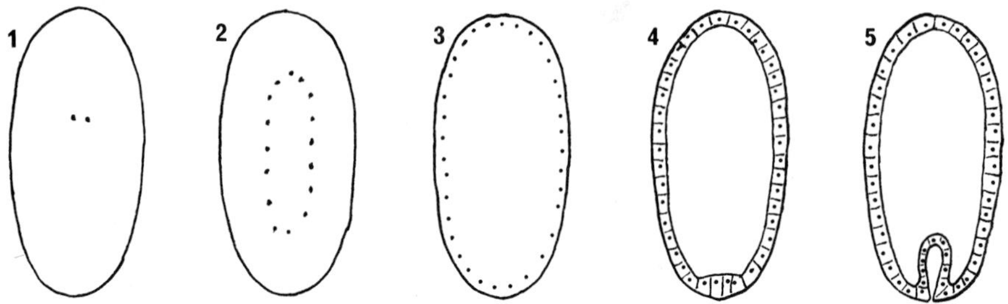

Figure 4.5. Early stages in insect development. 1, egg consisting of a delicate outer shell or "chorion" surrounding yolk, in which the zygote nucleus has just begun to divide. 2, numerous nuclear divisions follow, to produce a large number of "cleavage nuclei". 3, the cleavage nuclei migrate to the periphery. 4, membranes enclose each nucleus within a single cell, to produce a thin monolayer, the "blastoderm", a portion of which is thickened, forming the germ disc. 5, infolding of the germ disc produces the "germ band", which develops into the embryo.

In some organisms the whole body is coenocytic or contains polyploid nuclei and in others distinctive cell and tissue types may differ from one another in ploidy and numbers of nuclei per cell. Amongst some fungi, coenocytic hyphae become partitioned by diaphragms known as "septa" into compartments. Depending on the species and developmental stage of the fungus, these compartments may contain only a single nucleus or pairs of nuclei, or be multinucleate with variable numbers of nuclei. Commonly, the first hyphae to emerge from a spore are coenocytic, and only become partitioned subsequently. This situation is paralleled in insect embryos where the fertilized egg becomes multinucleate before it becomes multicellular (Fig. 4.5).

Another circumstance where mitosis is not accompanied by division of a cell into two equal daughters occurs when a change of scale accompanies division. Examples are provided by the production of spores, which are either smaller or larger than the "mother cell" from which they are derived, and by the production of some kinds of egg cells. These cases involve the dissociation of "offspring" from a parent body and are therefore commonly classified as asexual reproduction.

4.1.2. Changing Genetic Identity: Gene Transfer and Recombination

Whereas clonal proliferation maintains identical sets of genes within individual boundaries, there are various ways in which genes can transfer across these boundaries, causing new combinations and distribution patterns to arise.

For many people, the most familiar mechanism of genetic transfer involves the process of "fertilization" during sexual reproduction in eukaryotic organisms. This integrational process involves the incorporation into the same nucleus of single sets of genetic information derived from each of two mating partners. Nuclei which for this reason contain two sets of genes are known as "diploid", as are cells and individuals that contain such nuclei.

The events leading up to and following fertilization include a process known as "meiosis" which involves two rounds of nuclear division (Fig. 4.6). This process reduces (halves) the number of chromosomes from two sets in a single diploid nucleus to one set in each of four "haploid" daughter nuclei. Meiosis can be thought of as a kind of differentiation, not only because it involves subdivision, but also because it generates genetic variation.

During the first division of meiosis, pairs of chromosomes that have duplicated during a previous interphase line up on a microtubular spindle. Each pair contains one chromosome from each of the mating partners. The chromosomes are described as "homologous" because they contain equivalent kinds and amounts of genetic information. However, they can differ at particular individual sites or "loci" due to deletions, substitutions, duplications or inversions in the sequence of bases. As a result of these differences, individual genes may occur as two alternative versions or "alleles" within a diploid nucleus. Where the alleles at a particular locus on two homologous chromosomes differ, the nucleus, cell or individual containing these chromosomes is said to be heterozygous for these alleles; where they are the same it is said to be homozygous.

Once the homologous chromosomes have paired and lined up, they separate and draw (or are drawn) apart to opposite poles of the spindle, prior to becoming enclosed by nuclear membranes. An interphase follows during which there is no DNA replication, and then the second meiotic division occurs when sister chromatids separate on a spindle as in mitosis.

The special feature of the first division of meiosis is that it involves separation of *duplicated chromosomes* rather than *sister chromatids* and is followed directly by the second division rather than by DNA synthesis. This is how the number of chromosome sets is reduced from two to one.

Besides reducing chromosome number, the first division of meiosis also results in the *recombination* of alleles derived from the original mating partners. This process ensures that whilst each haploid daughter nucleus receives on average *half* its DNA complement from each parent, only very exceptionally will it receive the *same* half as one of its sisters. So long as their parents differ from one another genetically, meiotic sisters will also differ. Unlike those of mitosis, the products of meiotic divisions are therefore genetically variable.

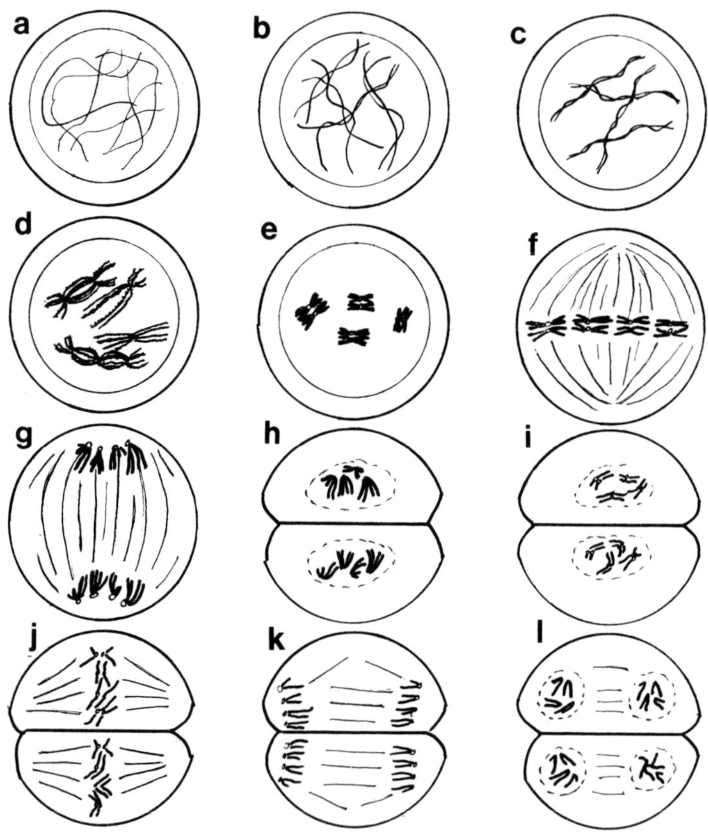

Figure 4.6. Meiosis: a, chromosomes begin to condense ("leptotene"); b, homologous chromosomes pair up with one another ("zygotene"); c, chromosomes thicken and form "chiasmata" or "crossovers" ("pachytene"); d, extensive RNA synthesis occurs, chromatids become distinct, and chromosomes appear to separate except at cross-over points ("diplotene"); e, chromosomes become very condensed ("diakinesis"); f, chromosomes line up at equator of the spindle ("metaphase I"); g, homologues separate ("anaphase I"); h, cells divide, nuclear membranes form ("telophase I"); i, prophase II; j, metaphase II; k, anaphase II; l, telophase II.

One of the ways in which recombination occurs during meiosis depends on the fact that each homologous chromosome has an equal chance of drawing to one or other pole of the meiotic spindle. Sets of genes on different chromosomes (non-homologues) therefore have the freedom to *assort independently* from one another. For example, if a diploid nucleus contains three sets of homologous pairs, represented as A1 and A2, B1 and B2, C1 and C2, eight possible combinations of non-homologues can occur in haploid daughter nuclei: A1B1C1, A2B1C1, A1B2C1, A2B2C1, A1B1C2, A2B1C2, A1B2C2 and A2B2C2. Generally, the total number of permutations is equal to 2^n, where n is the number of chromosome pairs. This number can be very large, e.g. 2^{23} in human beings where the number of chromosome pairs is twenty-three.

Recombination can also occur because the chromatids of homologous pairs can break and rejoin with one another to form structures known as "chiasmata" prior to drawing apart (Fig. 4.7). The genetic consequence of chiasma formation is known as "crossing over"—the reciprocal transfer of information from one homologous chromatid to the other. Because of crossing over, even genes that are "linked" on the same chromosome (see discussion of supergenes in Chapter 2) have some freedom to recombine, i.e. to dissociate and reassociate, during meiosis. The further apart the genes are on a chromosome, the more likely is a crossover event between them, and so the greater is their degree of freedom or capacity for independent assortment from one another. By the same token, the closer together they are, the less likely they are to dissociate.

When it comes to thinking about how fertilization and recombination affect the distribution of genetic information in natural populations of organisms, sex can be confusing! One important source of confusion lies in the fact that the relative prominence of haploid and diploid stages in life cycles, and the degree to which individuals of discrete gender can be recognised, vary in different kinds of organisms. This variation often causes scientific specialists who study different organisms to find themselves talking, sometimes crossly, at cross-purposes! Needless to say, problems of defining contextual boundaries and distinguishing between determinacy and indeterminacy figure prominently in this cross-talk.

In many (but not all) animals, including ourselves, generations of relatively discrete, diploid, male and female individuals occur. The haploid phase in the life cycles of these organisms only exists as "gametes"—egg and sperm—so that diploid parents appear to beget diploid offspring. Evolutionary discussions of sex amongst zoologists therefore often focus on the "cost" of males (which in a population containing a 1:1 sex ratio results in a 50% reduction in fitness) and the benefits of heterozygosity.

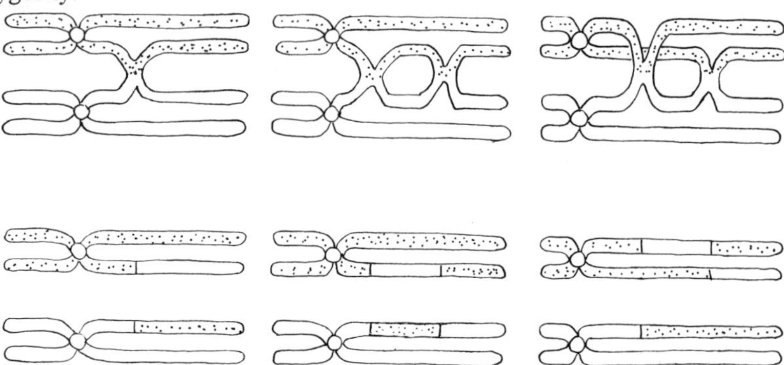

Figure 4.7. Chiasma formation and crossing over. Top row (left to right), single crossover involving two chromatids, double crossover involving two chromatids and double crossover involving three chromatids. Bottom row, the genetic consequences of the crossovers shown in the top row, once the chromosomes have separated.

By contrast, in many (but not all) plants and fungi, haploid phases may exist as independently growing "individuals" rather than just as gametes. Also, there may be no separation of gender, either because there are no identifiable male and female sex organs, or because the same individual can possess both kinds of sex organs. Often there is a clear cut *alternation of generations* between haploid and diploid phases (see Fig. 5.9), or even, as in some fungi where there is a long delay between mating and fertilization, between haploid and "heterokaryotic" phases (see Chapter 7).

What should be thought of as parents and offspring in these organisms? Are haploids to be regarded as the offspring of diploid parents, and diploids the consequence of mating between haploid parents? The same argument applied to animals such as human beings would imply that a father's daughter is really his granddaughter, the daughter of his sperm! Also, where the haploid generation consists of self-sustaining "individuals" subject to natural selection, then any argument about the significance of heterozygosity within individuals is irrelevant. Similarly, where there are not separate genders, then arguments about the 50% cost of males are also irrelevant. On the other hand, it may still be possible to argue from a restricted perspective that sex necessitates a 50% reduction in transmission of an individual's genes to the next generation due to having to involve a mate. This cost might be avoided by "self-fertilization", but fertilization between the gametes of a diploid "self" is equivalent to "inbreeding" between haploid "sister selfs".

All this can get very muddling! However, it does highlight the importance of defining contextual boundaries when discussing relationships between parents, offspring, inbreeding, outbreeding, selfing etc. It also emphasizes the need to be cautious when generalizing about the role of sexual reproduction in all organisms from knowledge only of a particular group.

Is it possible, nonetheless, to reach any generally applicable conclusions about the influence of sexual reproduction on the distribution of genes in natural populations? I will try to do so in the next few paragraphs.

To begin with, a fundamental consequence of sexual reproduction is that it results in *gene flow*. Whereas clonal proliferation confines genes to strict lines of descent that can be thought of as "genetic trajectories", sex allows genes to bridge across trajectories. A gene can therefore spread and become established far more rapidly via sexual transmission than by clonal descent. Whilst imposing constraints on individual freedom, sex is liberating for genes!

Depending on the degree of freedom of genetic transfer, the ultimate consequence of gene flow may be envisaged to be either the formation of *gene pools* or the formation of *genetic networks*. The concept of a gene pool basically assumes that genes have complete freedom to move around within a population boundary, so that their distribution amongst individuals is random. This assumption has been central to the development of the science of "population genetics", much of which is devoted to predicting changes of gene frequency between generations under various kinds of selection pressure. However, whilst it might make calculation easy, it is a thoroughly unrealistic assumption that ignores context and "averages out" variations in the distribution of individuals and groups of individuals.

In real populations there may be some *overall* limit to the distribution of genetic information, such as a geographical barrier. However, within this limit there is liable to be a complex mixture of constraints and opportunities for encounters between individuals. Real populations are therefore likely to become spatially and temporally structured into subpopulations of various sizes, with varing amounts of gene flow between them. Such spatially structured populations are known as "metapopulations". They are partially ordered genetic networks, not soup-like gene pools in which boundaries between individuals and groups are completely dissolved to leave genes as the "units of selection".

In genetic networks, genes are distributed *locally* amongst interbreeding individuals. No individual is capable of containing *all* the genetic information available in the population as a whole, and so each individual can be regarded as a differentiated component of the system, capable of occupying a distinctive niche. The source of all the genetic information is not recombination itself, but either comes from outside the system—by immigration—or by means of mutation; all that recombination can do is generate creative mixtures, like a dealer shuffling a pack of cards.

To summarize, sexual reproduction highlights the reciprocity between freedom of individuals and freedom for genes. Genes remain partitioned inside individual boundaries within sexual generations but are free to disperse and re-associate between generations. Individuals may be relatively free and subject to selection for much of their lives, but if they do not mate their genes will be lost unless there is also a mechanism for clonal proliferation. Whilst the maintenance of individual boundaries is therefore important, up to a point, for the survival of genes, so too is the integration of those boundaries during sexual reproduction. As open-ended networks, sexually interbreeding populations are both integrated and differentiated, with all the resilience, flexibility and predictable unpredictability that that implies.

Although sexual reproduction is usually regarded as a characteristic of eukaryotic organisms, genetic transfer mechanisms also occur in bacteria. Some of these mechanisms are reminiscent of sex; others indicate how genes can breach the boundaries even of very different organisms by means of "horizontal transfer".

The three known mechanisms of genetic transfer between bacteria are *conjugation*, *transduction* and *transformation* (Fig. 4.8). Conjugation involves a sex-like process in which one cell, often described as "male" acts as a donor, attaching by means of "sex pili" to a "female", recipient cell.

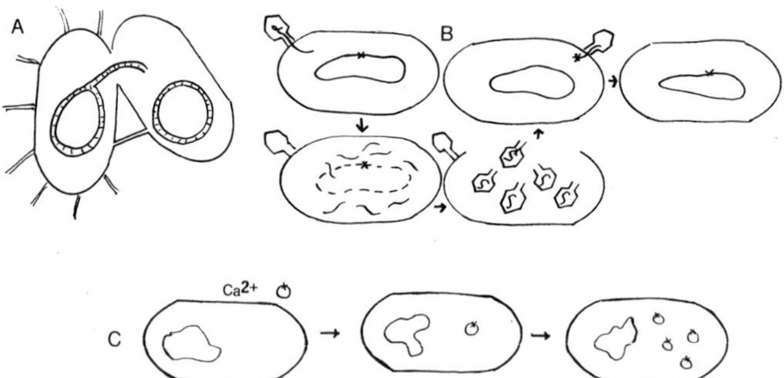

Figure 4.8. Mechanisms of genetic transfer in bacteria. A, "conjugation": a "male" cell with sex pili synthesizes a strand of DNA which it transfers into a "female" cell. B, "transduction": during an infection cycle, some viral DNA associates with a host DNA fragment (marked "x")—this fragment can then be transferred and become incorporated by recombination into a new host genome. C, "transformation", as commonly used in gene cloning, when a plasmid incorporating a particular DNA sequence (x) enters a host cell and begins to replicate autonomously.

Transduction involves the intervention of a third party, a virus or "bacteriophage", as the agent of transfer. When the DNA of such a virus enters a host bacterial cell, it can combine with some of the host DNA. This host DNA can then be transferred, along with the viral DNA, when the resulting virus particles infect other bacterial cells. The DNA of some bacteriophages is able not only to transfer host genes from cell to cell, but actually to integrate into and be replicated along with the bacterial DNA. This occurs in what are known as "lysogenic" cycles because they are only potentially cell-destroying ("lytic").

Transformation involves the uptake of "foreign" DNA by a bacterial cell from its immediate environment rather than directly from another cell or virus particle. In some bacteria this can occur relatively readily, but others have to be treated specially, e.g. by exposure to calcium ions at low temperature, before they will take up DNA in this way. Transformation is an important process in genetic engineering, representing one of the means by which DNA which has been manipulated outside cells so as to include particular foreign genes that can be incorporated into cells and then "cloned".

When transformation is used in genetic engineering, it often involves non-chromosomal pieces of DNA, which are normally circular, known as "plasmids". Plasmids contain an "origin of replication" (see Chapter 2) and so can multiply independently of the bacterial chromosome. Some of them are "transmissible" in that they can be transferred from cell to cell by means of conjugation. They are common inhabitants of bacterial cells, where they often carry genes conferring resistance to antibiotics, and have also been found in certain fungi.

One plasmid, the Ti (Tumour-inducing) plasmid in the bacterium *Agrobacterium tumefaciens*, has attracted particular interest, because it demonstrates how genes can be transferred between fundamentally different kinds of organisms. The bacterium causes a disease, "crown gall", that results in tumour-like growths at the junction between the roots and stem (the "crown") of many kinds of plants. The symptoms are due to the transfer of a region known as T-DNA from the Ti plasmid into the host plant cells, where it integrates with the host DNA. The T-DNA contains information which induces the host cells to proliferate, as well as to produce unusual amino acids, known as "opines", that serve as specific nutrients for the bacteria.

Here, then, is a case where a bacterium might be said to genetically engineer its host in a way that provides it with a unique natural habitat. If one such case exists, how many similar cases might await discovery? As I have mentioned elsewhere, there could be considerable opportunity over evolutionary time for horizontal genetic transfer between all kinds of organisms that become intimately associated in symbioses and close-knit communities.

The behaviour of the T-DNA in the Ti plasmid is characteristic of a general class of entities that are known as "mobile genetic elements" because they can relocate between and so rearrange different parts of a DNA molecule. Since many of these elements have no known "function" and appear to "hitchhike" on or piratize host DNA, they are sometimes referred to a "selfish DNA". They may form a large proportion of the repetitive DNA sequences that can be found in the genomes of eukaryotes.

Two important kinds of mobile elements are known as "insertion sequences", which do not contain genes, and "transposons", which do. Transposons have sometimes been referred to as "jumping genes", and in bacteria they often carry information that confers resistance to antibiotics. The movement of insertion sequences and transposons can activate and inactivate chromosomal genes, but since this movement appears to be random, it is not generally thought to represent a mechanism for ordering gene expression.

By contrast with those brought about by mobile elements, some kinds of DNA rearrangement are vital to successful patterns of cell differentiation. One of the most complex examples—and one which beautifully illustrates the creativity of association-dissociation interplay at the molecular level of organization—occurs in the vertebrate immune system.

The immune system is generally regarded as a "nonself recognition" system that prevents healthy animals from being overrun by invasive agents such as viruses, bacteria and fungi. It works by both detecting and eliminating these foreign agents.

This system has many parallels with sexually reproducing populations in that it can be thought of as an open-ended genetic network in which separate, genetically differentiated entities are distributed within a dynamic context. However, in the immune system, the context is defined by the body boundary of an individual multicellular animal, and the entities are cells.

Just one gene encoding a polypeptide about 100 amino acid units long is thought to have been the ancestor of the immune system. This gene has been duplicated and diversified over evolutionary time to give rise to a superfamily of genes and protein products of such versatility that it is capable of detecting a close to infinite range of "foreign" molecules.

Even artificially produced molecules will elicit an immune response. This implies that the system can operate without any previous exposure to the molecules in question during an organism's evolutionary past.

Correspondingly, invention precedes identification of the target and is built into the indeterminacy of a system that faces innumerable, unpredictable challenges during an organism's lifetime. Only inventiveness, not the inventions themselves, is passed genetically from one generation of organisms to the next. There is no competitive adaptation, but rather an intrinsic flexibility—a creativity within the system—that allows it to respond to rather than predict circumstances.

The flexibility of the immune system stems from the freedom of individual entities within immunity gene superfamilies to associate with one another in a multiplicity of combinations. These different combinations produce a multiplicity of protein products with different boundary configurations. It is therfore the freedom to recombine which results in the freedom from an infective burden that is implicit in the very term "immunity" (from the Latin, *immunus*, meaning "exempt").

The detective component of the immune system involves the binding of proteins, encoded by immunity genes, to specific molecular sites, known as "determinants" on the surface of an invasive agent or "antigen". The detective proteins are known as "antibodies" if they are secreted in solution by cells known as "B-lymphocytes", which, in mammals, originate in the bone marrow. If, on the other hand, the proteins are attached to the surface of cells known as "T-lymphocytes" originating in the "thymus gland", they are known as "T-cell receptors". Only antibody production will be discussed below; however, similar principles apply to the formation of T-cell receptors.

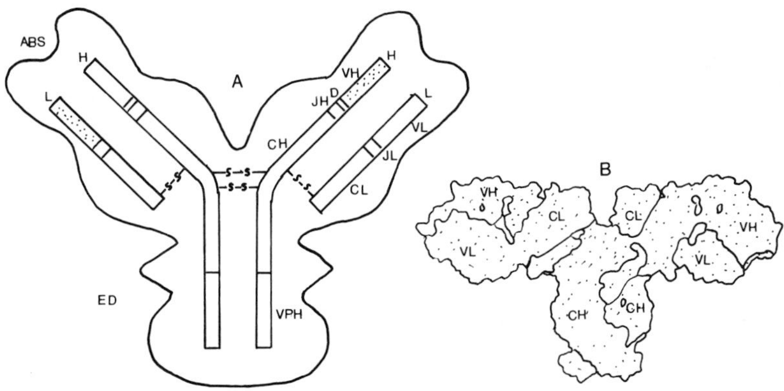

Figure 4.9. Antibody structure. A, schematic diagram showing the heavy (H) and light (L) chains joined by disulphide bonds and divided into sections as follows: heavy chain—variable (VH), diversity (D), joining (JH), constant (CH), portion varying between immunoglobulin classes (VPH); light chain—variable (VL), joining (JL), constant (CL); whole structure—antigen-binding site (ABS), effector domain (ED). B, outline of a three-dimensional model based on x-ray crystallography.

Antibodies are also known as "immunoglobulins"; they are Y-shaped molecules that consist of four polypeptide chains—two identical short or "light" chains and two identical longer or "heavy" chains (Fig. 4.9). Light chains are divided into two and heavy chains into four physical "domains" bridged by covalent "disulphide bonds". Disulphide bonds also link the four chains to one another. In mammals there are five classes of immunoglobulins—IgA, IgD, IgE, IgG and IgM—which have distinctive functions and differ from one another in the composition of their heavy chains.

Both heavy and light chains are segmented into regions which are either variable (V) in composition or relatively constant (C) linked by a joining (J) section. In heavy chains there is an additional diverse section (D) between the V and J regions.

As might be expected, it is the variable regions, located at the ends of the arms of the Y, which bind to antigen. The stem of the Y, comprising two constant heavy chain domains, serves as an "effector domain" that signals the initiation of processes able to eliminate antibody-bound antigen.

The amino acid sequence in the variable regions determines the configuration of the binding sites, which in turn determines the kind of molecule that can be bound. There is considerable specificity in antibody-antigen reactions so that even a single change of amino acid can alter the kind of antigen that can be bound.

Given such specificity and the enormous diversity of potential antigens, how can the equivalent diversity of antibodies be produced without necessitating thousands, if not millions of antibody-encoding genes?

The answer involves a remarkable brand of teamwork involving three basic kinds of "gene families", two of which prescribe two kinds of light chain (lambda and kappa) and the other of which prescribes the heavy chains. Each family occurs on a different chromosome and contains "libraries" or optional choices of segments encoding different kinds of V, J and D regions. For example, for the heavy chain in mice there may be around 300 V options, 20 J options and 4 J options.

In embryonic (undifferentiated) cells, all the component segments of a gene family are present stretched out and separated from one another along the length of the relevant chromosome. However, for a functional mRNA to be transcribed, it is necessary for a particular selection of V, J, D (for heavy chains) and C segments to be juxtaposed and joined with one another. This is achieved by rearrangement of the DNA (Fig. 4. 10).

A huge number of different rearrangements of the gene segments are possible. In the case of the mouse heavy chain there are 300 x 20 x 4 = 2.4 x 104 different combinations of V, J and D units. But that is only the beginning! The joining of V to

D and D to J regions is actually *imprecise*, such that a random few nucleotides are lost from the DNA at the point of joining, so changing the coding sequence. This increases the diversity of options by about tenfold. Not only are nucleotides lost from the joints, but a few nucleotides or "N regions" can actually be added, perhaps generating a further 100 or so possibilities, making a total of 2.4 x 10^7 for the heavy chain. For similar reasons there may be about 4 x 10^3 possible light chain combinations. Assuming that any light chain can combine with any heavy chain, the total number of combinatorial possibilities approaches 10^{11}, a very large number indeed. These rearrangements take place continuously during the proliferation and differentiation of B lymphocytes in the bone marrow of mammals.

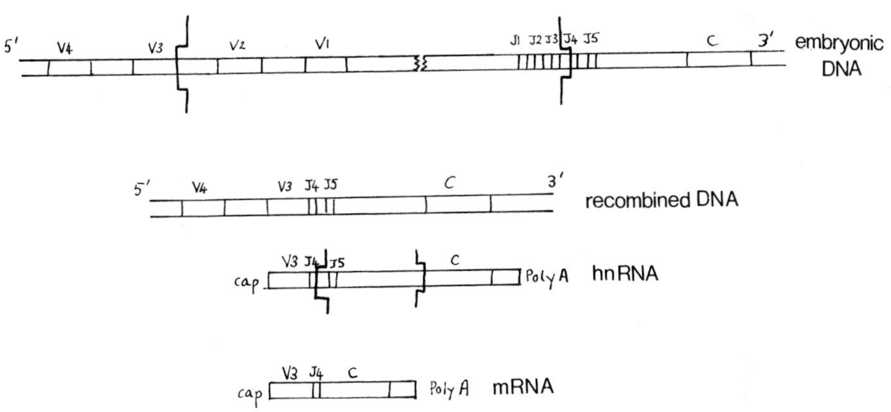

Figure 4.10. DNA rearrangement and RNA processing in an immunity gene family.

Once a lymphocyte has matured, i.e. got to the stage where it has acquired surface antibody, it ceases to proliferate and will in fact die after a few days unless it encounters the requisite antigen. Consequently a varied population of short-lived, individually unique B-lymphocytes is constantly being generated by the bone marrow, and is circulated through the body via lymph ducts and blood vessels.

Should a virgin lymphocyte encounter the requisite antigen, however, it becomes *activated*, sometimes with the assistance of "helper" T cells. It then begins to proliferate clonally, yielding a large number of antibody-secreting plasma cells that circulate through the body.

In addition to plasma cells, some of the products of activation are so-called "memory" cells. These cells are long-lived and provide a means of rapid response to an antigen that has previously been encountered during the life of the organism. An encounter between a memory cell and the antigen results in activation in the same way as with a virgin lymphocyte. However, the system does not have to await the appearance of the requisite lymphocyte in the same way as when the organism is first exposed. For this reason, adults are generally immune to many childhood diseases such as measles and mumps.

To summarize, each vertebrate individual inherits a set of genetic information from which a huge range of antibodies can be constructed by means of DNA rearrangements. These rearrangements are not predetermined by precise kinds or sequences of antigens, and effective antibodies can be produced even against molecules that have not previously been experienced by the organism or its ancestors. However, once an antigen has been detected, production of the appropriate antibodies is amplified by the proliferation of B lymphocyte clones, and the capacity to respond to subsequent exposures is reinforced by the long-lived memory cells.

It has been suggested that the antigens select the effective B lymphocytes in a classical Darwinian manner. However, no competitive struggle for existence is necessarily involved, and the antigens might more appropriately be regarded as niches for the lymphocytes, both defining and being defined by the configuration of the antibodies.

4.2. Opening, Closing and Extending Cell Boundaries

The proliferation, segregation and recombination of genes all take place within the context of chromosomes, plasmids, transposons, nuclei, mitochondria, chloroplasts, which in turn inhabit cells. I shall therefore now consider how cell boundaries are integrated and differentiated.

In Chapter 2, I stated that cell boundaries maintain a dynamic balance between expansive and constraining processes by allowing varying degrees of communication between the protoplasm and external environment. Critical to this balance are the deformability and permeability of the boundaries, which determine both the shapes assumed by cells and patterns of uptake, loss and throughput of resources.

4.2.1. Constraint Without Walls

Animal cells lack a cell wall and so are unable to prevent net losses through their cell membranes in resource-poor environments. This inability can be overcome by forming multicellular associations or by moving away from resource-poor regions. Alternatively an impermeable covering may be produced around individual cells or groups of cells.

In the absence of a wall, the shape of animal cells depends partly on their external surroundings and partly on the arrangement of a dynamic internal scaffolding made up of proteins known as the "cytoskeleton " (i.e. cell skeleton). These proteins consist of "microfilaments", "microtubules" and "intermediate filaments" respectively 7, 24 and 10 nm in diameter.

Microtubules are assembled by polymerisation and disassembled by depolymerisation of two kinds of protein subunits, "Â-tubulin" and "ß-tubulin". During assembly, pairs of Â and ß units associate in a helical array around a hollow centre. Microtubules provide the basis for the internal support and movement of whip-like cellular outgrowths, "cilia" and "flagella", as well as forming tracks along which the movement of cellular particles can be directed. As mentioned earlier, they also make up the structural framework of both mitotic and meiotic spindles.

Microfilaments, like microtubules, can both be assembled from and disassembled into protein subunits. The subunits in this case consist of "G-actin". Along with another filamentous protein, "myosin", they provide the means for contraction of muscle cells. They probably also bring about "cytokinesis" of animal cells, i.e. the separation into daughters during the final stage of mitosis by means of a constricting ring which pinches the cells apart.

In amoebae the association and dissociation of microfilaments is thought to underlie the transition between an outer gel-like layer of cytoplasm, "ectoplasm", just below the plasma membrane, and an inner liquid sol, "endoplasm", which is more mobile. The sol-gel transitions are thought to be influenced by hydrogen ions and calcium ions which respectively tend to increase and reduce chemical cross-linking of actin filaments. During "amoeboid movement", the protrusion of "pseudopodia" (false feet) is thought to involve driving endoplasm to a deforming tip where it sets, by cross-linking of actin, into ectoplasm. The process is therefore rather like some kinds of lava flow in which lateral boundaries solidify around a molten core. It is also reminiscent of the extension of some kinds of walled cells (see below).

4.2.2. Setting and Deforming Walls

Where cell walls are present, they fix the shape of the cells that they envelope. However, the cytoskeleton may still play an important role as temporary scaffolding that can readily be assembled, disassembled and relocated to sites where the wall is expanding. Here, the cytoskeleton may provide mechanical resistance to deformation before the wall has set and/or determine the pattern of delivery of wall components to the cell boundary. The interaction of hydrogen and calcium ions with actin filaments may therefore also affect the shape of walled cells.

Even if cytoskeletal elements and hydrogen and calcium ions do influence the shape of cell walls, the question remains as to how their own distribution pattern is regulated. Does this distribution pattern determine boundary configuration, is it determined by boundary configuration, or is there, as seems most likely, complex feedback between the two? The conundrum of system-boundary interplay returns to centre-stage.

As mentioned in Chapter 2, cell walls counteract unlimited expansion resulting from osmotic input of water following uptake of solutes by active transport. Correspondingly, they cause metabolically active cells to be pressurized when inhabiting environments where the external concentration of solutes is less than the internal concentration.

The fact that walled cells can become pressurized has some important repercussions. If the internal pressure is *exactly* counteracted by the resistance of the wall (plus any cytoskeletal or other kind of re-inforcement), then the cell's internal pressure is equal to the difference between the internal and external water potentials (see Chapter 2). If the cell has a stationary boundary, the rate of uptake of water is equal to the rate of leakage of water; the cell is described as "turgid" and its internal pressure is referred to as its "turgor pressure". However, for an expanding cell, where the boundary is "giving way", the situation is potentially far more complex.

If the internal pressure of an expanding cell is to be kept at a constant, "equilibrium" value, rates of net uptake (gross uptake minus leakage) must be increased in such a way as to match exactly the displacement of the boundary; i.e. net input = throughput = output (boundary displacement). If, however, the boundary gives way faster than can be supported by uptake, i.e. throughput exceeds input, the pressure in the system (and that deforming the boundary) will drop such that a "surge" is followed by a "lull". If, on the other hand, throughput becomes exceeded by input, pressure in the system will rise, such that a lull may be followed by some kind of surge. In other words, cells with deforming boundaries are liable to display nonlinear dynamics of the kind referred to earlier in this chapter.

The relationship between uptake, throughput and output, depends critically on three resistances. These are the resistances of the cell boundary to deformation and to passage of molecules and the resistance of the cell interior to displacement of cell contents.

Mechanisms which affect these resistances dictate the arrangement of cell boundaries in space and time. For example, the production and chemical cross-linking of microfibrillar compounds will increase rigidity, and the deposition of hydrophobic compounds will increase the impermeability of cell walls. Together, these rigidifying and impermeabilizing processes increase the "insulation" between the protoplasm and external environment and hence reduce energy transfer. There is increasing evidence that insulation is increased when, for whatever reason, the external environment becomes resource-depleted. This would make sense because it allows captured resources to be "sealed in", just as loft and wall cavity insulation reduces heat loss from our homes.

In elongated cell systems, the distribution of strengthening and sealant materials will not only affect the input of resources, but will also determine how this input leads to displacement of contents in a particular direction. This distribution will therefore be critical to changes in the form and activity of the systems.

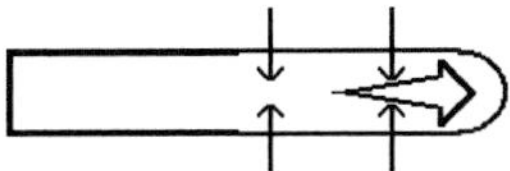

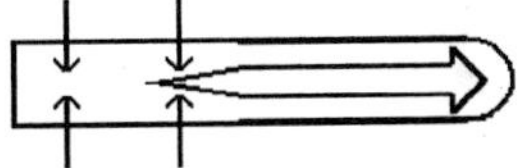

Figure 4.11. The effects of boundary deformability and boundary permeability on patterns of input (simple arrows) and throughput (tapering arrows) in elongated cell systems. Rigid boundaries are shown as straight lines, deformable boundaries as curves and impermeable boundaries as thicker lines. (Modified from Rayner et al., 1995).

In the first of two contrasting arrangements shown in Fig. 4.11, the most deformable part of the boundary occurs in a permeable region, whereas the rest of the boundary is relatively more rigid and sealed. Here, the extension of the permeable region that results from the cumulative displacement of fluid towards the deforming apex also proportionately increases the absorptive surface. Were this process to continue indefinitely, the rate of extension would increase exponentially towards infinity. However, in reality a threshold point is reached where the combined resistance to throughput exerted by the deforming apex and due to the diameter of the tube restricts displacement to a maximum rate, within a fixed length of tube. Any additional input can then only be accommodated by swelling or branching of the tube.

In the second arrangement depicted in Fig. 4.11, the most deformable region is associated with an impermeable boundary. Here, expansion depends on the distribution of materials from a region surrounded by a boundary that is permeable, and so receptive to input, but relatively less deformable.

So, in the first arrangement, assimilation and expansion occur in the same part of the system, whereas in the second arrangement, expansion occurs at a distance from assimilative sites. The first arrangement enables exploitation of energy-rich sites, whilst the second arrangement allows exploration across energy-poor regions.

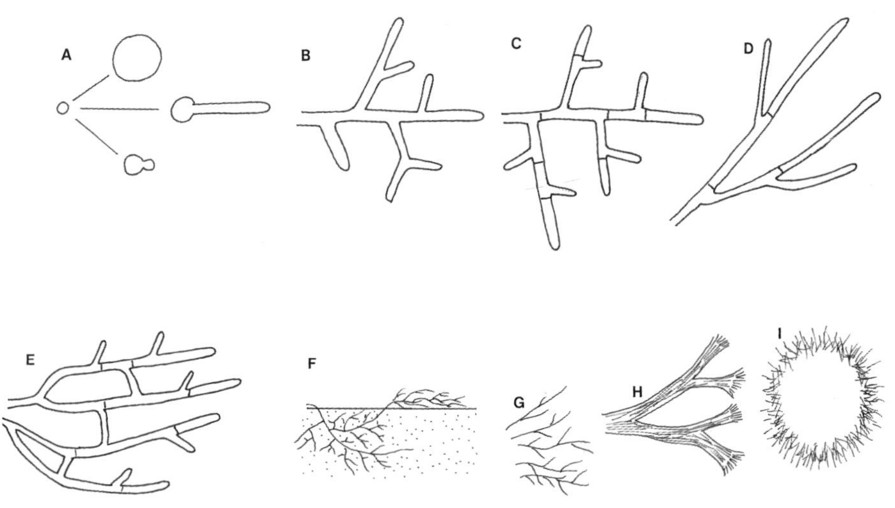

Figure 4.12. Organizational variation in mycelial fungi. A. A spore germinating to yield a germ tube, a giant cell or a budding yeast. B. A coenocytic (unpartitioned) hyphal branching system. C. A septate hyphal system with tributary-like branching. D. A septate system with distributary-like branching. E. An anastomosed (partially networked) hyphal branching system. F. Assimilative and non-assimilative hyphae (represented as single lines) growing respectively within and above the nutrient source (stippled). G. Diffusely organized hyphae. H. Hyphae organized into cable-like linear aggregates (mycelial cords). I. An annular mycelium (fairy ring) resulting from degeneration of the central region. (From Rayner, 1996a)

The mycelial systems of fungi appear to provide a good illustration of the dynamic relationship between exploitative and explorative states arising from different arrangements of relatively more or less permeable and deformable boundary regions. I have to say "appear to" because the relationship has not yet been proved and the growth patterns of mycelia have usually been explained as if these systems were simple additive assemblages of discrete, duplicating "growth units". However, there are many ways in which mycelia, or rather the way they are thought about, illustrate the challenges of moving from a conventional "building block" viewpoint to one based on dynamic boundaries.

Mycelia emerge from spores and then diversify into a wide range of alternative forms (Fig. 4.12). When a spore germinates by taking up water and nutrients, it often expands isotropically at first and then breaks symmetry (see Chapter 3), allowing one or more apically extending, protoplasm-filled hyphal tubes to emerge. Alternatively, a determinate developmental pattern can result in the formation of "giant cells" or yeast-like phases.

Once polarity has been established, the hyphal tubes may be fully coenocytic or they may become internally partitioned by inwardly growing cross-walls or septa. Sooner or later, the tubes branch, either in a tributary-like or a delta-like pattern. The branches may diverge or they may converge and fuse (anastomose), so converting the initially radiate system at least partially into a network. Whilst some parts of the system may be intimately associated with the nutrient source, others may become sealed off or emerge beyond the immediate sites of assimilation. The branches may remain diffuse or they may aggregate to form protective, reproductive or migratory structures. The latter consist of cable-like arrays and can often extend much faster than individual hyphae. Whilst some parts of an established mycelial system may continue to expand, others may stop growing and degenerative processes may set in.

Many studies have shown that the rate of increase in length of a hypha is exponential at first but sooner or later becomes linear. These observations have been used to support additive growth models. However, the phenotypic patterns capable of being produced by purely additive systems are strongly constrained, and could not include many of those shown in Fig. 4.12. They are physiologically constrained because the possession of a relatively open boundary allows uptake in energy rich fields, but also renders them highly dissipative (leaky) so that they discharge in energy poor fields. Purely additive mycelia would therefore be unable to explore spatial domains lacking nutrients or to survive temporal shortages. They are also geometrically constrained because of limitations in their pattern of branching. Purely assimilative systems branch in tributary-like patterns because the sites of uptake coincide with the sites of outgrowth. This pattern of branching connects resistances to displacement in series, and prevents the space-filling needed for the system to sustain an evenly growing margin that expands at a linear rate in more than one spatial dimension (a commonly observed feature in mycelia).

These constraints can be overcome by three kinds of dissipation-minimizing processes: anastomosis, boundary impermeabilization and autodegeneration. Anastomosis (fusion of branches) converts an in series system at least partially into a parallel-distributing network. It therefore increases throughput capacity, restores symmetry and enables amplification of organizational scale by allowing enhanced delivery to sites of emergence of, for example, rhizomorphs and fruit bodies. Boundary-impermeabilization converts the mycelium into a structure able either to conserve resources within rigid boundaries and so to survive temporary shortages, or to distribute resources to deformable boundaries and so explore across energy-poor fields. Distributive structures have fan- or delta-like branching, because the sites of uptake are remote from the sites of outgrowth. By integrating exploration and assimilation it is possible for a mycelium to sustain an evenly growing margin in more than one dimension. Autodegeneration enables a system to redistribute resources from redundant to actively expanding components (see Chapter 5).

It may therefore be of great significance that fungi with septate (internally partitioned) hyphae, and correspondingly impeded throughput, are also those in which the branches are most prone to anastomose with one another. These fungi can also greatly increase their operational scale, through the emergence of hyphae wider in diameter than their predecessors or of cable-like aggregations. The formation of these cables is analogous, both structurally and functionally, to the establishment of international telephone links once national networks are in place. They enable a mycelium to link separate assimilative sites over what can be a very long range. The rhizomorphic fungus, *Armillaria bulbosa*, for example is on record as having produced an individual network some 15 ha in area and estimated to weigh 100 tonnes and to be 1500 years old in a Canadian forest. Even larger networks are suspected in the forests of Washington State. The increased power supply through networks is also critical to another fundamental property of fungi with septate hyphae; the ability to build large fruit bodies from which they broadcast their spores to the outside world. Such are the consequences of being able to differentiate and integrate at the cellular level of organisation.

4.3. Intercellular Partitioning and Communication

As mentioned in Chapter 2, it is debatable whether fungal mycelia should be regarded as multicellular because although they can be compartmented by septa, the latter allow considerable protoplasmic communication. In some sense, therefore, mycelia represent the most extreme result of the diversification and integration of a single cell, and the *Armillaria* systems mentioned earlier could then be regarded as the largest, most complex *cells*, if not amongst the largest and oldest *organisms* in the world!

In other organisms, more clearly discrete cells either gather together to form cell societies, or proliferate from a single source within a common boundary to produce true multicellular individuals (Chapter 2).

Cell societies form when cells that have been proliferating as dispersed populations in resource-rich environments begin to associate with one another as their environment becomes restrictive. This situation parallels that of a mycelium which proliferates branches as a relatively uninsulated structure in nutrient rich domain, but becomes more coherent, due to dissipation-minimizing processes, in nutrient-poor locations. However, whereas the mycelium assumes a more ordered form by constraining activities within a variably sealed physical cell envelope, cell societies can also form as a result of the mutual attraction of their constituents.

A classical example of a cell society founded on mutual attraction is provided by the cellular slime moulds (Chapter 2). When populations of the amoeba-like cells of these organisms start to run out of food (in the form of bacteria), certain cells become attraction centres, causing the other cells to start to stream towards them. They do this by sending out pulses of a chemical stimulus, in the form of cyclic adenosine monophosphate (cAMP). Other cells respond to this stimulus by sending out a pulse of their own and then moving towards the original signal source. A strong positive feedback results, causing the cells to establish and to follow gradients of the cAMP.

The cellular slime moulds therefore exemplify one of the most important integrative mechanisms in social collectives, i.e. the production of and response to an attractive chemical that is capable of diffusing over long range. Such stimuli are generally known as "pheromones" and in many cases they also provide for a potent means of attraction between sexual partners (Chapter 7).

As mentioned earlier in this chapter and in Chapter 1, all kinds of multicellular structures, by whatever route they are produced, are potentially very unstable if the boundaries between adjacent cells remain complete. This is because of the disequilibria that can develop between neighbours and resulting autocatalytic redistribution "to those that hath from those that hath not".

Such instability can be prevented by the formation of communicating junctions that allow at least some degree of protoplasmic continuity between cells (Fig. 4. 13). These communicating junctions, which parallel the septal perforations and anastomoses of fungal mycelia, can take a variety of forms.

In some animals, structures known as "gap junctions" contain sets of protein tubes, called "connexons", of sufficient diameter to allow sugars, amino acids or other similarly sized molecules to pass from one cell to the next. Gap junctions can form in minutes, by assembly of preformed connexon subunits, and can also break off communication within seconds. Interestingly, malignant cancer cells do not readily form gap junctions with their neighbours.

The cytoplasmic connections between adjacent *Chlamydomonas*-like cells in *Volvox* were mentioned in Chapter 2. A more widespread kind of cytoplasmic connection between cells of multicellular plants is known as a "plasmodesma". Plant cells, being surrounded by a wall cannot divide off from one another in the way that animal cells can—through the production of a constricting ring of actin microfilaments. Rather a so-called "cell plate" spreads out from the midpoint between dividing daughters. In this situation, plasmodesmata appear to form from strands of endoplasmic reticulum (Chapter 2) that traverse the cell plate. However plasmodesmata are also capable of forming, as do hyphal anastomoses in fungi, by breaching pre-existing walls between neighbouring, non-daughter cells. Neighbouring cells within a plant tissue can thereby become linked into a network.

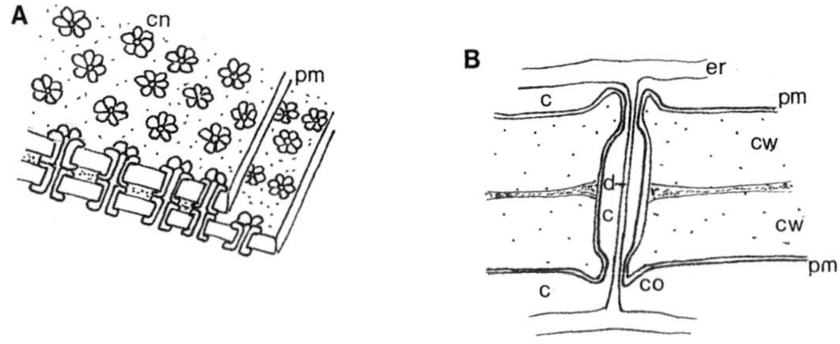

Figure 4.13. Communicating junctions between cells. A, gap junction; B, plasmodesma; c, cytoplasm; cn connexon; co, collar; cw, cell wall; d, desmotubule; er, endoplasmic reticulum; pm, plasma membrane. (Not to scale)

4.4. Defining Tissues

As mentioned in Chapter 2, truly multicellular organisms don't just have differentiated cells (as in a sponge)—they also have distinctive tissues that fulfil varied roles in the life of the organism.

Tissues are assembled from distinctive arrays of cells which form following the proliferation of more or less similar or identical cells that ultimately originate from a single progenitor. Where this progenitor is a "zygote", i.e. the product of sexual fertilization, the set of cells that arises from it comprises what can be called an "embryo".

At first, most embryos consist of little more than a group of more or less similar cells. For tissues to form, some kind of re-organisation has to occur, allowing the cells to become heterogeneously distributed into distinctive subdomains where they follow different developmental pathways. The way that this re-organisation occurs contrasts markedly between most higher plants and animals, presumably reflecting the relative indeterminacy and determinacy of these organisms.

Basically, in higher plants, the embryo becomes polarized and new cells are added at the apices of the resultant elongated structure either by proliferation of a single apical cell, or of sets of dividing cells known as "meristems". The apical meristems occur at the tips of shoots and roots and are responsible for the production of cells which give rise to all the tissues of what is known as the "primary plant body". In woody plants, secondary lateral meristems known as "cambia" then give rise to the conductive tissues within bark ("phloem") and wood ("xylem"), thickening the roots and stems in the process. The localization of cell division within apical meristems also occurs in colonial Cnidaria and Bryozoa (Chapter 2) and is a basic feature of indeterminate multicellular structures, analogous to the extending tips of hyphae and other cellular filaments.

By contrast, in the majority of animal embryos the production of new cells for the body as a whole occurs within all the developing organs and tissues and so is not localized. The development of these embryos is usually classified into a series of stages that follow on from one another in a fixed sequence.

Following fertilization of an animal egg cell by a male gamete, rounds of mitosis accompanied or followed (as in insects) by *protoplasmic cleavage* cause cell and/or nuclear numbers to increase rapidly—in a series of doublings. The nutrients fuelling this increase are derived entirely from the egg and so the divisions are generally not accompanied by expansion of the early embryo.

In some cases the egg is subdivided more or less evenly and gives rise to a body of cells known as a "morula"—as happens in the embryos of mammals. In other cases, the presence of yolk at the "bottom" of the egg impedes cleavage, so that many more cells are formed at the top. In the most extreme cases, as in birds, division is confined to a relatively small area overlying the yolk.

As division continues, an internal space or "blastocoel" develops. This either converts the structure into a hollow ball of cells, a "blastula", or, as in birds, separates a layer of cells, the "blastoderm" from the underlying yolk (Fig, 4.14).

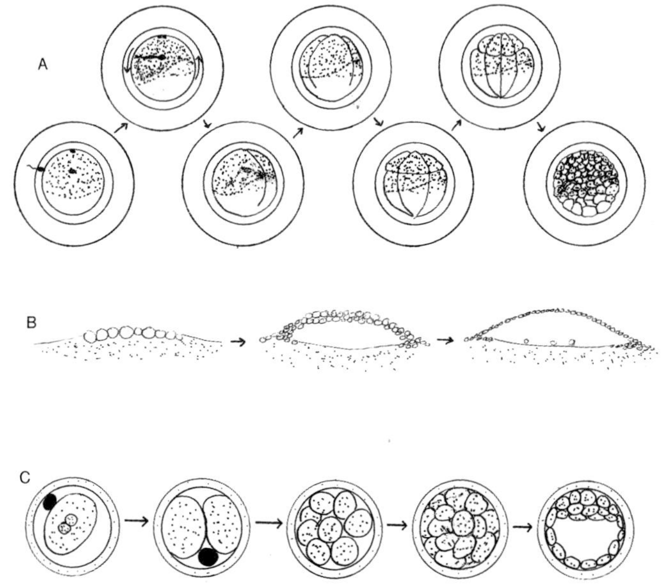

Figure 4.14. Blastulation in vertebrate embryos. A, in frogs, the egg is enclosed within a vitelline membrane and jelly coat, and divided into a dark-pigmented "animal hemisphere" and a light-pigmented, yolky, "vegetal hemisphere"; meiotic breakdown products known as "polar bodies" occur at the animal pole; entry of the sperm into the animal hemisphere is followed by a rotation of the "egg cortex" as shown, leading to the formation opposite the sperm track of a "grey crescent"; the first cleavage is vertical and usually bisects the egg into right and left halves; the second cleavage is at right-angles to the first and usually separates front (ventral) and back (dorsal) halves; the third cleavage is transverse; further cleavages transform the egg into a hollow ball of cells, the blastula, with smaller cells in the animal hemisphere. B, in chicks, cleavage is confined to a small area overlying the yolk (stippled) and results in the formation of a "blastoderm" overarching a subgerminal cavity—the central region ("*area pellucida*") of the blastoderm appears transparent when viewed from above and is surrounded by a thicker, more opaque "*area opaca*". C, in mammals, the egg is surrounded by a "*zona pellucida*" which also encloses a polar body; fusion of the egg and sperm-derived "pronuclei" is followed by cleavages which double the number of cells until a "morula" is formed (fourth stage shown); finally a blastoderm is formed surrounding a cavity or "blastocoel" and also enclosing an inner cell mass (shown in section) which will develop into the embryo itself.

The next stage of development, "gastrulation", is critical to the definition of boundaries between embryonic tissues (Fig. 4.15). It involves massed movements of cells in such a way that they become organized into three "germ layers" known as "ectoderm", "mesoderm" and "endoderm". In hollow blastulas, gastrulation involves progressive invasion of the blastocoel. In birds, it involves the movement of cells down through a groove abutted by two ridges, the "primitive streak", so that they come to underlie the original blastoderm.

Gastrulation involves both individual and collective movements of cells. Individual cells move in an amoeba-like manner and follow particular pathways through the embryo. How these pathways are defined is uncertain, but embryonic cells have been shown in the laboratory to follow lines scratched in the bottom of culture dishes. Once the cells reach the end of whatever lines they are following, they stop and adhere to their neighbours, apparently in a very specific way, and enter diverse developmental pathways.

The three germ layers formed during gastrulation ultimately give rise to very different kinds of tissues. In vertebrates, the outermost layer, or ectoderm gives rise to the nervous system and outer layers of skin. The endoderm forms the lining of the gut and digestive glands. The mesoderm forms the internal skeleton, muscles and organs other than those associated with digestion. Eventually, the mesoderm becomes divided by another cavity, the "coelom", within which the internal organs are suspended, supported and separated off from one another by sheets of tissue known as "mesenteries".

The processes that follow gastrulation and result in the formation of the tissues and organs of the mature organism clearly have to be orchestrated in a very precise sequence if "monstrosities" are to be avoided. It is generally considered that this sequence is administered by a genetic programme that involves the controlled switching on and off of distinctive sets of genes, enabling the cells and tissues to specialize in different roles (i.e. to become differentiated). For this programme to operate successfully, it is important for the developing embryo to be buffered, as far as is possible, from the effects of a variable external environment. With some important exceptions (e.g. gender determination in reptiles), and in marked contrast to indeterminate forms, the programme is therefore followed without reference to external conditions. Whilst feedback mechanisms are important, they operate internally, within and between the developing cells and tissues, rather than with the embryo's surroundings. However, if these surroundings are allowed to impinge on development, for example due to exposure to a drug or virus, they can have disruptive effects.

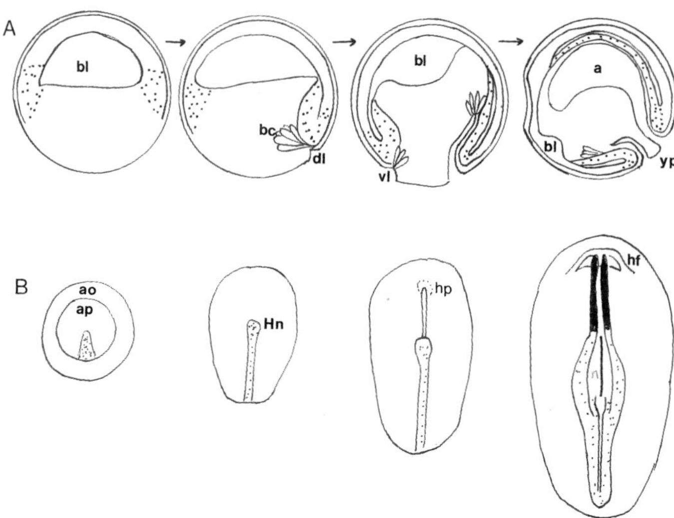

Figure 4.15. Gastrulation. A, progression from blastula to late gastrula in *Xenopus*, the African clawed toad, as seen in median section: a belt of tissue, the "marginal zone" (stippled) drives the process, which starts at a depression, the "dorsal lip of the blastopore" (dl); this depression expands to form a full circle, the blastopore, at the other side of which the "ventral lip" (vl) forms; invagination occurs all round the blastopore and from the dorsal lip results in the formation of a cavity, the "archenteron" (a); invagination is associated with the formation of "bottle cells" (bc); as the archenteron forms, "involution" of the marginal zone leads to a general shift of tissue in a dorsal direction; the blastopore shrinks and at the late stage is occluded by a yolk plug (yp); the blastocoel (bl) is progressively obliterated. B, gastrulation in a chick embryo, as seen from above: the "primitive streak" (stippled) has "Hensen's node" at its tip, from which the main body of the embryo is generated, beginning with the formation of a "head process" (hp), followed by formation of a "head fold" (hf), neural folds (shown black) and "notochord" (n). (After Slack, 1991)

The expectation that genes set the agenda for embryonic development of determinate body forms can therefore readily be appreciated. However, it remains to be understood how this agenda actually comes to be followed so strictly.

The mechanisms which are thought to be important in setting the course of embryonic development are known as "determination", "embryonic induction" and "positional information".

Determination involves the progressive narrowing down of the developmental options that a cell or cell line can follow towards fully differentiated end-products. The fertilized egg cell itself is "totipotent", capable of giving rise in its progeny to all possible options. However, as blastulation, gastrulation and subsequent steps proceed, this totipotency declines, and the cells make "irrevocable decisions". The whole process has been likened to a ball rolling down a hillside or "epigenetic landscape" dissected by a bifurcating series of valleys (Fig. 4.16), even though real drainage patterns actually branch in the opposite direction to this (see Chapter 3). As the ball rolls and follows one fork or another, its ability to change course declines. Anastomoses between the valleys are not included in the picture.

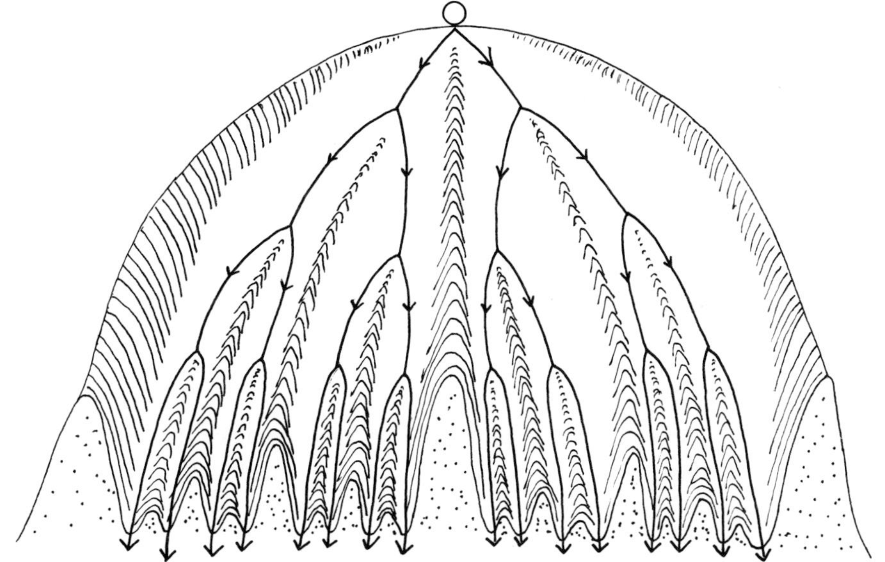

Figure 4.16. Developmental decision-making, of the kind envisaged by the concept of the epigenetic landscape.

The process of determination in animal embryos can be traced by means of transplantation. Cells are either isolated or relocated experimentally and their respective ability then to give rise to whole animals or to differentiate in accord with their new surroundings is monitored. Determined cells cannot do either of the latter.

Here, a fundamental difference between determinately developing animals and indeterminately developing plants should be noted. With few exceptions, determined animal cells cannot change their developmental course. However, even fully differentiated plant cells can, so long as they remain alive, regenerate into whole individuals. In the laboratory, such regeneration involves "de-differentiation" into an unco-ordinated mass, known as "callus". Under appropriate circumstances some of the callus cells can become co-ordinated in local centres to give rise to roots, shoots or indeed whole embryos. This is of considerable practical importance as it enables plants to be propagated, or "cloned" by means of "tissue culture".

Determination occurs at different stages in the embryos of different animals, and may be related to how quickly heterogeneity becomes established. In some cases, heterogeneity exists even within the egg cytoplasm and appears to be responsible for early restriction of developmental potential.

Sea urchin eggs, for example, are divided into a less refractile (light scattering) half, the "animal pole" and a more refractile "vegetal pole". The first two cytoplasmic cleavages are longitudinal with respect to these poles, and the individual cells can be separated and give rise to whole animals. However, the third cytoplasmic cleavage is equatorial and so separates two groups of four cells with different cytoplasmic contents. The animal half can be separated, and will continue to grow, but develop mostly ectodermal tissues. The vegetal half on the other hand develops mostly endodermal and mesodermal tissues. In frog eggs, a pigmented "grey crescent" is distributed around the equator. If the first cleavage bisects the grey crescent, both the resulting daughter cells can be separated and develop into tadpoles. If the division is parallel to the grey crescent, only the cell containing the crescent is capable of producing a tadpole. By contrast, in mammals, where the egg cytoplasm is more homogeneous, cells can be isolated even at the eight or sixteen cell stage and give rise to complete embryos.

Whilst some developmental paths may be predetermined in this way, by cytoplasmic ingredients from the mother, subsequent determination is thought to depend on processes of embryonic induction. These processes increase in importance as the embryo becomes progressively more heterogeneous. They involve the use of neighbouring dissimilar cells, generally originating from a different germ layer, as "reference points" that somehow bring about determination in a particular cell type.

A good example of embryonic induction is provided by the synthesis of skin proteins in vertebrates. Skin consists of two layers, an "epidermis" derived from ectoderm overlying a "dermis" derived from mesoderm. The kinds of structures produced by the epidermis, such as hair, feathers or scales involves the synthesis of particular kinds of proteins known as "keratins", and depends on the underlying dermis. So, if, for example, dermal cells from the leg of a chick embryo are transplanted under the epidermis overlying a wing bud, the epidermis is induced to form feathers of a type usually formed on the leg. However, if leg epidermis is transferred to the foot, it gives rise to scales, not feathers. The same epidermal cells are therefore capable of giving rise either to feathers or to scales depending on the inductive stimulus they receive from dermal cells that have already been determined.

It is obviously important during development that cells should not only be capable of differentiating into appropriate types, but that they do so in appropriate arrangements. An arm should develop as an arm and not as a leg.

The ability to develop in appropriate arrangements can partly be explained if induction occurs progressively, in a graded sequence. An example of progressive induction has been shown to occur in vertebrate limbs.

Vertebrate limbs develop from "buds" which emerge from the flank of the embryo and consist of a cap of ectoderm overlying a core of mesoderm. Removal of the ectoderm cap of a chick wing bud at an early stage prevents morphogenesis of the wing beyond the elbow, but at a later stage may prevent morphogenesis beyond the wrist. Also, if the tip of an old limb bud is transplanted to the stump of a young wing bud, the wing develops without its middle elements, whereas the reverse experiment results in duplication of wing components.

It appears likely, then, that the apical cap acts as an inducer of determination, such that cells laid down after brief exposure to the cap form upper limb components, and those laid down later form lower limb components.

There is also evidence that a region at the rear of the limb bud determines the number of digits, perhaps by producing a diffusible influence or "morphogen". It is thought that cells whose content of this morphogen is above a certain threshold are induced to synthesize more, whilst those with a content above a yet higher threshold destroy the morphogen. A counteractive embryonic "field" is thereby set up in which groups of cells amplify or suppress the influence in response to the productivity of their neighbours. Spatial oscillations in concentration result which determine where bone tissue is to form.

As a result of variation in morphogen concentration, by the time a cell ends up in its final position in the limb it either has acquired or then acquires a unique "positional value". The positional information shared amongst neighbouring cells can then provide reference points that allow a damaged or even (as in some amphibians) amputated limb to be repaired or regenerated in an appropriate pattern.

Even if the idea that variation in morphogen concentrations determines where bone tissue will form is correct, the question arises as to what bone-producing processes actually involve. The probable answer to this question provides yet another illustration of the importance of generating paths of least resistance in determining patterns of association and dissociation of cells. The limb bud can be thought of as an arena, bounded from the external environment by ectoderm. The external environment of the cells within this arena consists of a jelly-like material, known as the "extracellular matrix" which spaces the cells apart from one another. Any substance which causes the matrix to liquify will allow the cells to draw closer together to form a more condensed aggregate. One such substance is the enzyme "hyaluronidase" which degrades a principal component of the extracellular matrix, hyaluronic acid. Secretion of this enzyme is therefore thought to initiate the condensation of cells that leads to the differentiation of bone tissue.

It is generally assumed that differentiation, the process that follows determination, involves specification of a particular range of gene products, so the question arises as to how such specification could be achieved. One possibility would entail actual changes in the genetic information content of the cells, for example by amplification, deletion or recombination.

Whilst there is relatively little evidence for genetic changes other than in special cases like the immune system (see earlier), there is evidence that the information content of many differentiated cells remains intact. For example, the capacity for de-differentiation of plant cells clearly suggests that the totipotency of their nuclei is retained. Furthermore, classical experiments with frog embryos have shown that nuclei can be transferred from determined cell types to egg cells whose nuclei have been destroyed, and give rise to normal embryos.

It therefore seems most likely that determination and subsequent differentiation are in general achieved "epigenetically", by changes in gene expression rather than in gene composition. Such changes imply that sets of genes can somehow be switched on or off according to requirements.

An important recent breakthrough in understanding one of the ways in which such changes could be brought about has involved the discovery of "homeotic genes". These genes were first discovered in the fruit fly, *Drosophila*, where mutations in them can have such curious effects as inducing a leg to develop in place of an antenna. A number of these genes have been cloned and shown all to contain a "consensus" sequence, specifying similar sets of 60 amino acids (cf. Chapter 2). Similar sequences, or "homeoboxes" have been found in frogs, mammals and fungi. It is thought that the homeobox sequence may be characteristic of regulatory proteins, known as "transcription factors" that are capable of binding directly to DNA and thereby promoting or inhibiting gene expression.

Even more recently, the homeotic genes in *Drosophila* have been found to operate at the end of an hierarchical cascade of gene expression that provides successive layers of positional information that is used to specify distinctive developmental regions. The first of these layers is provided by maternal genes which are expressed in cells surrounding the egg and give rise to head-to-bottom and front-to-back gradients of protein morphogens known as *dorsal* and *bicoid*. These gradients regulate the expression of "gap genes", glorifying in such names as *hunchback* and *buttonhead*, which regulate two tiers of "pair-rule" genes that influence the segmentation of the embryo. The pair-rule genes then regulate the "segment polarity" genes which define the front and back boundaries of individual body segments.

The fact that this cascade of gene expression begins in maternal cells establishes an important point. This is that from the outset of development there is no absolute boundary between the generations that requires the embryo to develop solely in accord with its own genetic information content.

4.5. Interconnecting Tissues: Nervous and Vascular Infrastructures

According to the combine harvester principle, the separation of distinctive life-maintaining functions into local domains or tissues with specialized attributes allows each function to be achieved with maximum efficiency and minimum interference. However, the very process that leads to differentiation also enhances the interdependence between increasingly self-insufficient specialisms.

For combines to operate, it is vital to have in place some form of re-integrational infrastructure, in the form of interconnective conduits or "pipelines", that allows transmission of resources and information between specialisms. These pipelines either conduct liquid, as in the "vascular" systems of plants and animals, or electricity, as in the nervous systems of animals.

The pattern of development of nervous and vascular infrastructures is fundamentally indeterminate, resembling that of a foraging mycelium as they connect up their sites of supply and discharge (cf. Figs 3.3 and 3.4). Like the hyphal tubes of a mycelium they may be variably partitioned and cross-linked, and their lateral boundaries are variably insulated so as to receive and distribute input with minimal dissipation.

Nervous systems contain two types of cells: elongated "neurons" transmit electrical impulses and variously shaped "glia" provide "packaging" around the neurons. The neurons are often bundled together into cable-like structures known as nerves.

Neurons can take a variety of forms, but, in vertebrates, are commonly divided into four distinct regions (Fig. 4.17). The "cell body" contains a nucleus and most of the ribosomes and endoplasmic reticulum; it is the site of synthesis of protein and membrane components that are transported to other parts of the neuron. Leading into the cell body is a gathering-system of branched "dendrites" which receive nervous inputs from other neurons and integrate them. A distributive channel, or "axon" leads out of the cell body *via* a junction known as the "axon hillock". During development axons can both branch and elongate at their tips. In so doing, they maintain and proliferate connections at specialized junctions known as "synapses", both with tissues (notably muscle and gland cells) and with other neurons.

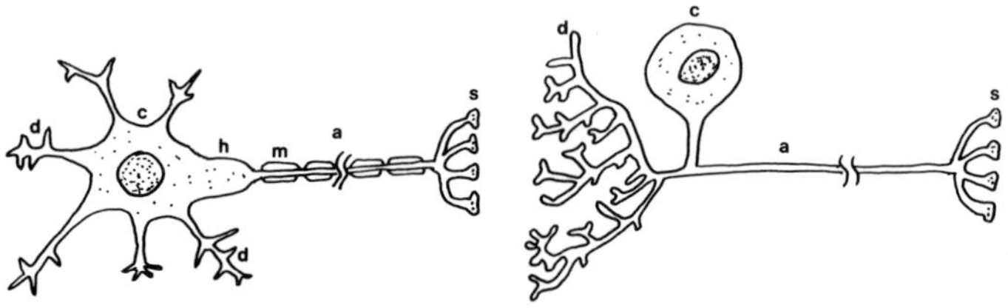

Figure 4.17. General features of vertebrate (left) and invertebrate (right) neurons: a, axon; c, cell body; d, dendrites; h, axon hillock; m, myelin sheath; s, presynaptic terminals. (Based on drawings supplied by Stuart Reynolds)

Patterns of production and transmission of electrical impulses through nervous systems have been shown to depend on counteractive processes. These processes are based on the differential permeability of nerve cell membranes to electrically charged ions, notably positively charged "cations" of potassium (K^+), sodium (Na^+) and calcium (Ca^{2+}), and negatively charged "anions" of chloride (Cl^-).

As with many other living cells, an electrical potential difference or "resting potential" is maintained across the plasma membrane of nerve cells because the membrane behaves like a potassium electrode. This is due to potassium being much more concentrated inside the cell than it is outside, and the membrane being much more permeable, at rest, to potassium ions than it is to other ions. The tendency of potassium ions to diffuse down their concentration gradient out of the cell is counterbalanced by the negative charge inside the cell generated by the ions that leave. This process is very economical and doesn't consume energy unless the potassium concentration gradient is allowed to run down, whence it must be restored by active transport processes. These processes pump Na^+ out into the external solution in exchange for K^+ (see Chapter 2). The net result is that the interior of neurons becomes negatively charged relative to the exterior and the membrane is said to be "polarized".

Nerve impulses are initiated when the membrane becomes locally depolarized, such that the interior of a neuron starts to become positively charged relative to the external solution. This depolarization causes certain "voltage-dependent" Na^+ channels in the membrane first to open, allowing Na^+ ions to flow into the cell down their concentration gradients, and then to close. The entry of the Na^+ ions has the effect of causing more voltage-dependent Na^+ channels to open. An autocatalytic feedback results which may continue until all the sodium channels are opened: the action potential is therefore an "all-or-none" event. As this happens, positive charge becomes displaced along the nerve, so initiating a "travelling wave" or pulse of depolarization, the "action potential", which spreads along the neuron.

Both the development and passage of nerve impulses have much in common with the spread of a "Mexican wave" through a football crowd, where spectators take cues from their neighbours first to stand up and raise their arms and then to sit down. Whether an action potential is produced depends on the counteractive effect of repolarization due to the opening of voltage-dependent K^+ channels which is initiated shortly after the opening of Na^+ channels. Repolarization prevents an action potential from being attained if the initial depolarization is below a critical threshold, or it results in the restoration of the resting potential after the action potential has passed (Fig. 4.18).

In line with the principles discussed earlier with respect to fungal hyphae, the degree and rate of spread of charge along a neuron depends on two properties, the *permeability* of the cell boundary to ions and the *conductivity* of the cell interior. The more permeable (i.e. the less well insulated) the boundary, the more readily charge will be dissipated by leakage, causing the depolarization to diminish unless it can be taken up and re-amplified by voltage-dependent Na^+ channels. The greater the cross-sectional diameter of the neuron, the more ions there will be per unit length to carry the current, and the further on average they can move before being discharged across the membrane.

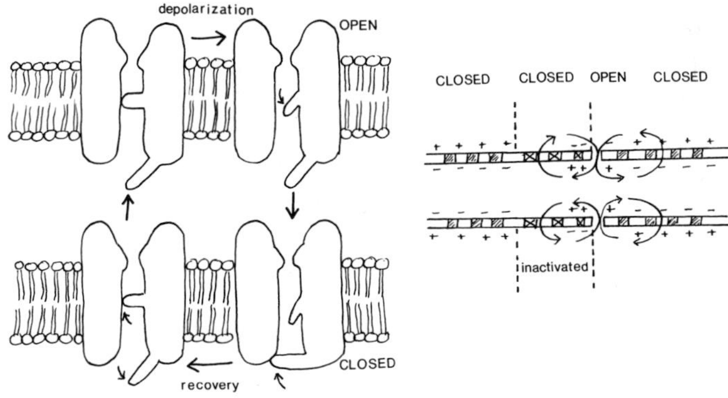

Figure 4.18. How an action potential is generated. Right, diagram illustrating depolarization and passage of an action potential from left to right, associated with opening and inactivation (x) of sodium channels—arrows indicate influx of sodium ions and efflux of potassium ions. Left, model of molecular mechanism of opening, closing and inactivation of sodium channels.

Wide, well-insulated neurons therefore conduct a nerve impulse further, faster and more efficiently than narrow, uninsulated ones. It is therefore understandable that the distributive components of neurons, i.e. axons, are both wide and insulated.

In vertebrates, insulation is provided by specialized glial cells, known as "Schwann cells" or "oligodendrocytes" that wrap their specialized plasma membranes around individual axons to form a many layered coating known as "myelin" (Fig. 4.19). The crippling disease, multiple sclerosis, results from loss of myelin from parts of the brain and spinal cord.

Myelinated axons are only exposed to external solution at junctions between the glial cells, known as "nodes of Ranvier". Depolarization can spread without attenuation and at very high speeds between these nodes, whilst the presence of voltage-dependent Na^+ channels at the nodes allows an action potential to be relayed.

An important feature of action potentials is that once initiated, they travel unidirectionally. This can partly be understood in terms of the fact that Na^+ channels upstream of an action potential cannot be re-opened by spread of depolarization until the resting potential has been restored. However, this only explains how unidirectionality is maintained, not how it originates. Here, some asymmetry, or means of breaking symmetry, around the initiation site may be needed. Such asymmetry is related to changes in the properties of neuronal membranes at the junctions where action potentials are characteristically initiated. These junctions include axon hillocks and synapses.

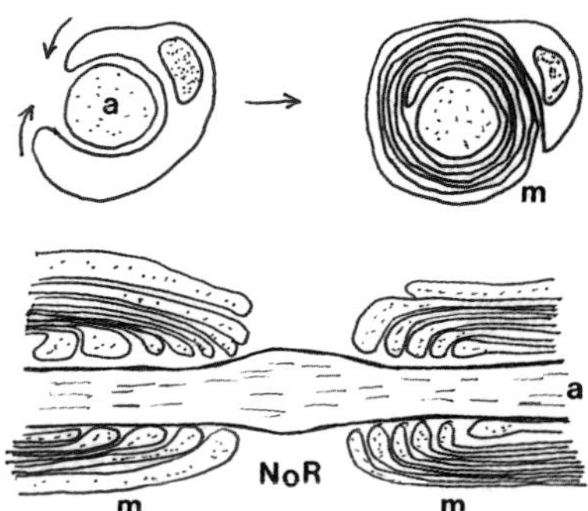

Figure 4.19. Insulating nerves. Top drawings show how Schwann cell membranes wrap around an axon (a) to form myelin sheaths (m); bottom drawing shows a longitudinal section through a node of Ranvier (NoR). (Based on drawings supplied by S.E. Reynolds)

Synapses are sites where nerve impulses are either relayed or impeded, again depending on the balance between counteractive processes. Some synapses are "electrical", consisting of gap junctions that relay an action potential without delay. More commonly, synapses are chemical (Fig. 4.20).

Chemical synapses involve the release of "neurotransmitters" that diffuse across the "synaptic cleft" and bind to receptors on the downstream (postsynaptic) cell. Here arrival of an action potential from the upstream (presynaptic) axon triggers a rise in cytosolic Ca^{2+}, via the opening of voltage-dependent Ca^{2+} channels. This rise then triggers the exocytosis of membrane-bound vesicles, which release the neurotransmitter chemicals. Binding of some neurotransmitters, of which the best known is acetylcholine, induces depolarization of the postsynaptic cell membrane, and hence transmission of an action potential. Other neurotransmitters, like gamma amino butyric acid (GABA) trigger "hyperpolarization" and so impede development of an action potential.

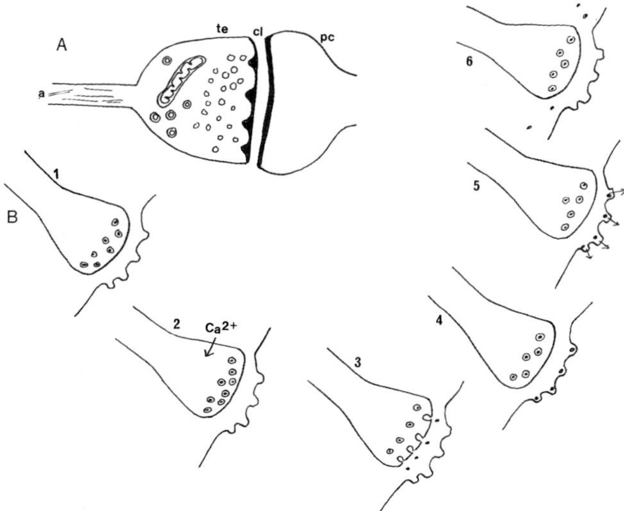

Figure 4.20. How chemical synapses work. A, basic structure of synapse in a longitudinal section: an axon (a) containing microtubules and neurofilaments leads into a terminal (te) containing mitochondria and coated and uncoated vesicles, followed by a cleft (cl) abutted by characteristic thickenings and the postsynaptic cell (pc). B, sequence of events (numbered 1-6) resulting in passage of an action potential across an excitatory synapse: 1, action potential arrives; 2, calcium channels open; 3, neurotransmitter released from vesicles, probably by exocytosis; 4, transmitter interacts with receptors; 5, activated receptors open ion channels; 6, transmitter dissipates. (Based on drawings supplied by S.E. Reynolds).

Depending on the identity of the neurotransmitter, synapses may be "excitatory" or "inhibitory", respectively propagating or resisting an action potential. A single neuron may receive impulses from up to thousands of other neurons that synapse with it. Whether such a neuron fires an action potential will then depend on the overall balance between inhibitory and excitatory signals that it receives.

Whereas neurons conduct electrical current, other kinds of infrastructure, e.g. the blood vessels and lymph ducts of animals and the xylem and phloem of plants, channel fluid flows. As with nervous infrastructures, these systems contain both gathering and distributive components.

Higher plants typically consist of two complementary, potentially competitive, but ultimately interconnected and interdependent systems—roots and shoots. Root systems consist both of highly branched, short-lived, absorptive components ("short roots") and indefinitely extending, explorative and conductive components ("long roots"). The absorptive roots gather and the conductive roots distribute soil solution. Shoot systems likewise consist both of indefinitely extending axes which explore the aerial environment and assimilative offshoots (typically, leaves) which harvest sunlight by means of photosynthesis. The products of photosynthesis, in the form of organic compounds or "assimilates", are partly allocated to further proliferation of the shoot system and partly distributed to the roots via an intervening trunk or stem.

Mineral nutrients and water are transported through a set of pipelines organized into a tissue known as xylem, and transport assimilates through another set of pipelines, within a tissue known as phloem. In herbaceous plants and in young shoots and leaves, xylem and phloem are associated with one another in cable-like "vascular bundles" or veins. In perennial plants, the xylem is normally contained in a central cylinder of wood, whilst the phloem is a component of bark. The external boundary of bark consists of an insulating layer of dead cells impregnated with a hydrophobic, corky substance, known as "suberin".

Wood is a very heterogeneous tissue, containing several kinds of cells, many of which lack protoplasm and have walls impregnated with a hydrophobic, polyphenolic compound, lignin. The cells are not completely isolated from one another, but interconnect by means of openings known as pits. Two basic kinds of conducting systems occur in wood—an axial (lengthwise) system of "tracheids" and (in flowering plants) "vessels", and a radial (horizontal) system of "medullary rays" (Fig. 4. 21). Water is both driven through wood by means of pressure generated osmotically in the roots ("root pressure") and also, more significantly, is pulled through by hydraulic tension. The latter results from evaporation of water ("transpiration") from the leaves.

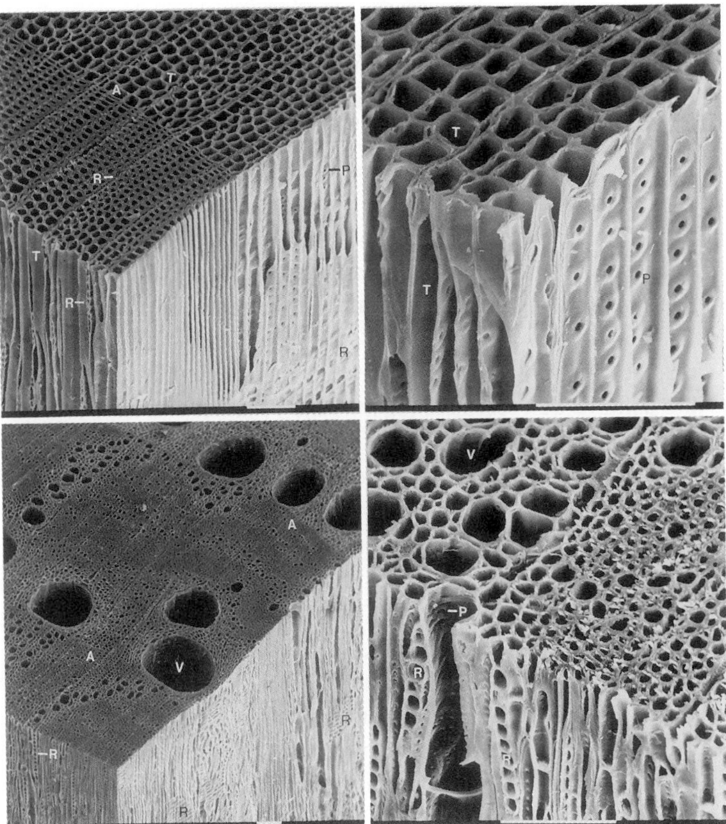

Figure 4.21. The three-dimensional structure of pine (top photographs) and oak wood, as revealed by the scanning electron microscope. A, junctions between earlywood and latewood, defining the boundaries between annual rings; P, pits —in pine these are "bordered", having a distinctive rim; R, medullary rays; T, tracheids; V, vessels. Scale bars represent 100 μm (Photographs kindly supplied by M.P. Ansell).

Phloem is also a heterogeneous tissue containing several cell types. The conductive component consists of "sieve tubes" or "sieve cells", so called because of the perforations in their end and/or side walls which connect one cell to the next in the series. Alongside the sieve tubes are "companion cells" which load the sieve tubes with assimilate at sugar-rich sites. Transport through sieve tubes is generally considered to be due to the development of osmotic gradients that cause bulk flow of solution from sugar-rich sites to locations where sugars are being depleted.

Transport between animal tissues is also often effected by a combination of gathering and distributive systems of branched pipelines. The distributive systems consist of arteries, through which blood is pumped by a heart. The gathering systems return circulatory fluids to the heart and consist of both veins and lymph ducts. Veins characteristically contain one-way valves that prevent backflow.

Veins and arteries are surrounded by relatively impermeable layers of muscle and connective tissue which are especially thick in arteries. However, at their gathering and distributive end-points, these major blood vessels characteristically branch into progressively finer sets of thin-walled, permeable channels known as "capillaries". The capillary systems enable oxygen, carbon dioxide, nutrients and waste products to be transferred between the tissues and blood stream and are organized into patterns that resemble the intricate venation of a leaf.

Tissues or groups of cells that are not tapped in to a capillary supply cannot proliferate. The expansion of capillary systems so that they maintain and/or establish connections with the tissues that they supply is therefore an important component of embryonic development. However, at later stages in the life of a determinate organism, the proliferation of capillary systems may also hasten the onset of processes resulting in disease and death. This situation arises when the capillaries establish connections with solid tumours, so releasing the latter from an energy-deprived state and allowing their cells to multiply.

The processes resulting in the "vascularization" of solid tumours demonstrate the classic themes of indeterminate development—polarization, branching and anastomosis—only too clearly. They begin with a response of the "endothelial cells" that form the wall of blood vessels to a proteinaceous chemical compound known as "tumour angiogenesis factor" produced by the tumour. This compound diffuses into the tissues and extracellular matrix surrounding the tumour and activates endothelial cells in neighbouring blood vessels to produce proteases and collagenases. These enzymes create channels or "capillary sprouts" in the extracellular matrix that the endothelial cells move into and proliferate within. Initially, the sprouts arising from a parent blood vessel grow out more or less parallel to one another. However, they then start to converge and fuse to form loops or anastomoses. From the primary loops, new sprouts emerge and the process is repeated until the tumour is penetrated.

Interestingly, there is one very important group of multicellular animals, the arthropods, which does not circulate blood through veins and arteries. Instead, these animals have an internal blood cavity or "haemocoel" that contains a pool of blood which bathes the tissues.

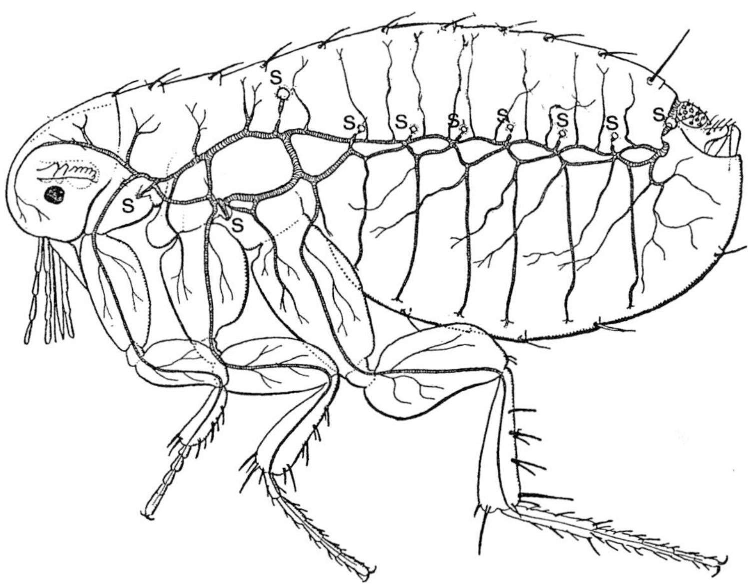

Figure 4.22. Insect airways: one half of the tracheal system and spiracles (s) of a flea. (After Wigglesworth).

The bodies of arthropods are encased by a relatively undeformable external cuticle. In terrestrial arthropods, such as many insects, this cuticle is often relatively impermeable but contains openings, "spiracles" which lead into a series of progressively more finely branched air channels. These air channels, or "tracheae", connect the internal tissues with the outside environment (Fig. 4.22) and so remove the need for the blood system to circulate respiratory gases (oxygen and carbon dioxide). An analogous system of proliferating air channels occurs in the lungs of terrestrial vertebrates (Fig. 4. 23). However, the localization of lung tissue to one part of the body requires that it is connected to a circulatory blood system which gathers and distributes respiratory gases from and around the remainder of the body.

As well as providing routes for the passage of resources and waste products, circulatory systems provide channels for long-range communication between tissues, in particular by means of chemical signals known as "hormones". Hormones may be defined as chemical compounds that are produced in one part of an organism's body and have an effect on cell behaviour or development at another part (the "target" site).

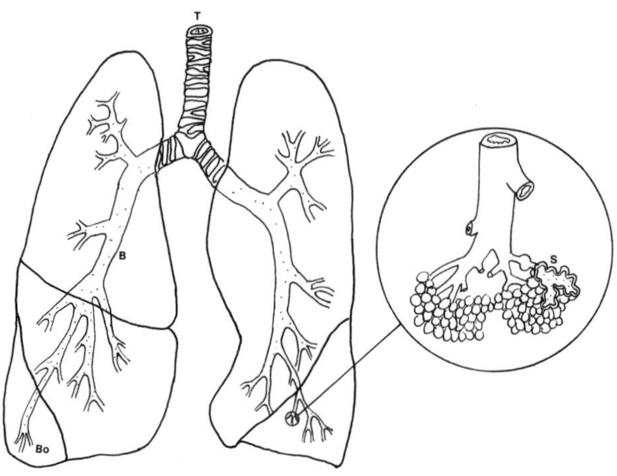

Figure 4.23. Diagram of human lungs illustrating progressive subdivision from the windpipe or "trachea" (T) into "bronchi" (B) and "bronchioles" (Bo), culminating in "alveolar sacs" (S).

In animals, hormones are usually produced in specialized tissues or organs, and can either be hydrophobic (water insoluble) or hydrophilic (water-soluble). The binding of hydrophilic hormones to receptor molecules on target cell surfaces, triggers a cascade of events that usually leads rapidly to a change in cell activity. A well known example is the "fight or flight" hormone, adrenaline. Hydrophobic hormones, on the other hand, characteristically diffuse across the plasma membrane and interact with protein receptors in the cytoplasm or nucleus, leading to a change in gene expression (mRNA synthesis). The steroids or "sex hormones" and the product of the thyroid gland, "thyroxine", are examples.

Plant hormones, of which six main kinds are known (indole acetic acid or "auxin", the "cytokinins", ethylene, abscisic acid, gibberellic acid and jasmonic acid) all affect some aspect of growth or development. Their modes of action are less well understood than those of animal hormones. This may reflect the fact that although plant hormones operate within an indeterminate context, researchers have expected them to involve similar mechanisms to animal hormones.

The principal effects of many plant hormones seem to involve changes in cell boundary properties and internal metabolism. For example, auxin plasticizes the cell wall, so making it more extensible when subjected to hydraulic pressure resulting from the uptake of water. Auxin may at least partly be distributed through incipient xylem tissue and has been shown to result in lignification of the walls of cells through which it is conducted. It may therefore have a "self-reinforcing" or "canalizing" effect on its own distribution.

As might be expected, there is a complex, counteractive interplay between the plant hormones. This interplay results in varied patterns of shoot and root emergence, extension and branching as well as various kinds of sealing off and redistributional processes that will be described in more detail in Chapter 5.

4.6. Structuring Societies and Communities

Many of the organizational principles upon which social structures are based reflect themes, and elaborate upon mechanisms that occur in cells and tissues. As organisms proliferate in unrestrictive, energy-rich regimes, they dissociate into highly subdivided, dissipative, competitive organizations that are unsustainable in the absence of continual replenishment of resources. However, as conditions become locally or generally restrictive due to external impediments of some kind or as a consequence of resource-depletion by the organisms themselves, non-dissipative organizations become favoured. From this stage onwards, counteraction between dissociative and associative trends determines and is determined by variations of resource supply from the environment and leads to increasingly complex and heterogeneous organizational patterns. Limitations on dissociation allow mass movements to be channeled in particular directions and provide scope for co-operation through the opening of communication channels.

The opening of communication channels generates the networking infrastructures of the most complex societies and communities, and provides the basis for these organizations to amplify their operational scales. However, these infrastructures also have the potential to take on a "life of their own", enslaving and even draining those individuals that once they might have served. As I will discuss further in Chapter 8, there are many signs that some of the infrastructures of human societies have done and are doing just this.

Both the form of social infrastructures and the nature of the material that they transmit can vary immensely. As mentioned in Chapter 2, the indeterminate mycelial systems of fungi can serve to interconnect living plants and recycle their remains in terrestrial plant communities. This raises the issue of whether it is the superstructure of the plants or the infrastructure of mycelia that regulates activities in these communities. In an analogous fashion, the infrastructure provided by natural waterways and their valleys have profoundly influenced the development of human societies.

In some societies, such as those of ants and cellular slime moulds, chemical signals or "pheromones" modulate organizational and behavioural patterns. In animals with the appropriate sense organs, sight, sound and touch all play important roles. Human beings have re-inforced natural passageways with road, canal and rail systems and achieved the ultimate, but costly, freedom of flight for transport both of themselves and resources. They have also aped, in their pipelines, power lines and cables, the water- and electricity-conducting systems of fungal mycelia, blood vessels, vascular bundles and nerves.

Human beings have also used their society infrastructures (media) to communicate with one another by means of various kinds of language. As methods for the recording, retrieval and transmission of words and symbols have advanced, so language has become of predominant importance in accelerating changes in human societies. These changes have been much much more rapid than could be accomplished by alterations in the genetic programming of individuals. Indeed, it has been suggested that communication-based human social evolution has now largely become uncoupled from human biological evolution, leaving us, as individual animals, adrift and unable to cope with the unpredictable dynamic systems that we have created.

For there can be no doubt about it; drawing on the energy assimilated by individual human beings, our infrastructures are evolving in the characteristic patterns of indeterminate systems. Within these infrastructures units of intellectual information, encoded in words or symbols, and referred to by some as "memes" to indicate their analogy with genes, emerge locally, proliferate, compete, network and fade. Languages, formulated over millenia in diverse cultural settings, compete for access to international networks which cannot operate in the presence of language barriers or incompatibilities. The heterogeneity of thought and culture that has evolved with the diversification of language may get lost on information superhighways.

This may be both "good" and "bad" news. It may be good news because communication is vital to mutual understanding. It may be bad news because loss of barriers means loss of control—as when the control rods are removed from a nuclear reactor allowing a chain reaction to run towards catastrophe. The loss of diversity of language may also imply a loss of diversity in thinking. Language is a niche for thoughts and hence provides the opportunity that enables ideas to be communicated and the dynamic boundaries within which those ideas can be formulated.

4.7. Conceptual Frameworks—Deductive and Inductive Reasoning

One of the more worrying implications of the predominance of language in regulating human societies is its association with one particular type of thinking—analytical thinking. This association may be an inevitable consequence of the location of the speech centre within the left hemisphere of the brain.

Analytical thinking is the kind of thinking, if it can be called that, which revels in discreteness, precision and separation into components that can be examined without reference to others in a step-by-step sequence. It is very easy to communicate and to teach: all that needs to be done is to resolve a system into its components by applying predictable step-by-step procedures formulated in words or symbols. In such ways, the particular can be deduced from the general—if A applies, then B follows. Errors can be eliminated, along with imagination. "Progress" is ensured by the natural selection of criticism. Argumentative word games and legalistic power struggles are the basis for "discussion". There is "right" and there is "wrong".

There is no doubt that this kind of intellectual differentiation has its place. However, that it should be considered to be the predominant and even the most respected form of "brain power" is another matter.

For reasons already discussed, analytical reasoning can only ever refine and particularize; it cannot generalize or innovate. The latter ends can, however, be accomplished by inductive reasoning, a kind of parallel processing which enables general conclusions to be derived from particular observations. However, it is very difficult to describe in words and symbols and in a step-by-step manner. Those who try to describe the outcome of inductive reasoning are frequently accused of "woolliness", and, worse still, of not producing "testable hypotheses". Such criticisms signify a lack of understanding of interconnectedness and a failure to perceive that although testable hypotheses need in themselves to be narrowly prescribed, they may be derived from the generalizations supplied by inductive reasoning.

Darwin's theory of evolution by natural selection is a classic example of the outcome of inductive reasoning. The theory itself may not have been testable as a whole, but deductions from it have certainly spawned a wealth of analytical research. By the same token, having initially provided an opportunity for new avenues of thought, there are now signs that Darwin's theory has become constraining. Some of its boundaries may therefore need to be reshaped or even abandoned if there is to be further intellectual development.

CHAPTER 5

VERSATILITY AND DEGENERACY

1. Reshaping and Abandoning Boundaries: Getting Out of Ruts

1.1. Entrenchment

Both in the long and in the short term, and for varying reasons, living systems are prone to get stuck in their ways.

On the one hand, processes of differentiation and competitive refinement produce increasingly specialized states from which any radical departure is liable to be unadaptive in the short term. Also, although successive rounds of differentiation into specialisms can produce more and more fine scale variation, this is due to proliferation of detail rather than the emergence of fundamentally new life patterns which depends on integration.

On the other hand, integration and enhanced communication can result in the continual reinforcement and reiteration of existing patterns by enabling the retention and efficient cycling of gathered resources. Once in a hole it really can become difficult to stop digging and set off in a new direction!

1.2. New Horizons

Getting stuck need not matter as long as nothing changes. However, changelessness is not a feature of a biosphere populated by dynamic living system boundaries. Heterogeneity (non-uniformity) abounds, both in space and in time. Life-supporting resources are supplied and depleted in variable patterns. System-boundary interplay continually generates new niches over an enormous range of scales. Attributes which have proved adaptive in some environmental settings cease to be so in others. Both in the long and in the short term, prosperity depends on being *versatile*—able to reshape physical or behavioural contextual boundaries in response to new or changing demands or roles.

1.3. Breaking Free

The formation of an "establishment" resists any tendency of elements at the boundary of a system to differentiate or integrate in response to new horizons. If this establishment is allowed to become too powerful, all capacity for constructive change is lost; the system may then stagnate and slowly degenerate.

Some means of escape from this fate by separating off from old contextual boundaries is therefore vital to any versatile system. The process of separation may involve actual detachment, in which case it may be perceived as an act of severance or reproduction. Alternatively, it may involve some means of partitioning one phase from another.

1.4. Redistribution

Reproduction, breaking free by means of detachment, allows dispersal to distant locations in space and/or time, but the extent to which resources from the "parent organisation" or establishment can be used to support its offspring is restricted.

On the other hand, if something like a one-way valve is inserted into an integrated system, the establishment can continue to contribute to new developments. Also, once a component at the boundary of a networked system has become a sink, receiving resources, then it may benefit from an enhanced power supply which allows new avenues of opportunity to emerge (see Chapter 4).

The basic principle of achieving redistribution by means of partitioning involves relocating boundaries in such a way that the establishment becomes a part of an emergent offshoot's external environment and hence a source of resources (Fig. 5.1). Redistribution therefore turns to advantage, and may even enhance the degenerative processes which would in any case ultimately ensue in a non-versatile system. In versatile systems, degeneracy provides means for recycling and renewal rather than demise. Death, the abandonment of old contextual boundaries, becomes a way of life.

Degeneracy and versatility are therefore interconnected. Remarkably, there is also increasing evidence that at least at the cellular level of organization they involve the same kind of chemistry. This chemistry is based on the reactivity of a molecule that might well be described as the world's first dangerously addictive drug—a molecule that most of us living, breathing creatures regard as the very essence of life but which is likely to kill us in the end!

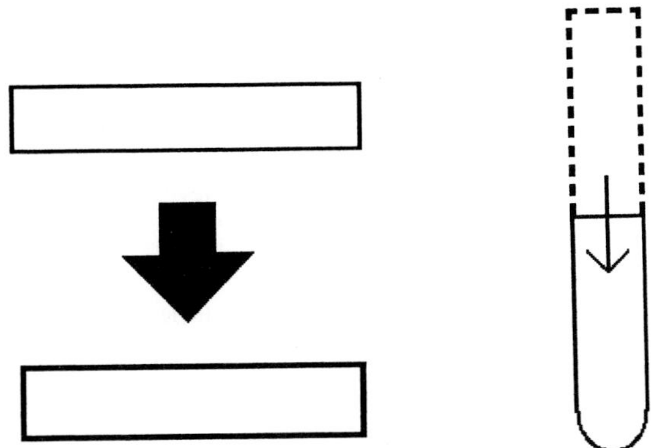

Figure 5.1. The contrast between energy-conservation in a system with a fixed boundary and redistribution in a system with a dynamic boundary. Energy conservation (left) involves the conversion of an already rigid (represented by a straight line), but permeable (represented by a thin line) boundary into an impermeable (represented by a thick line) boundary. Redistribution (right) involves the internal partitioning of a system such that actively deforming (represented by a curved line) components can be supplied with resources (as indicated by an arrow) from fixed components with a degenerating boundary (represented by a broken line). (Modified from Rayner et al., 1995).

The molecule that probably most fundamentally connects versatility to degeneracy is none other than oxygen. Although we tend to take the high proportion (about one fifth) of oxygen that occurs in earth's atmosphere for granted, the situation was very different when life forms first began to emerge on the planet. In fact, there was probably very little, if any oxygen in the atmosphere before the process known as photosynthesis really got going. Then, as more and more oxygen became available, a major opportunity and crisis loomed.

Oxygen is capable of unleashing large amounts of chemical energy. However, this energy can be deployed both constructively and extremely destructively, depending on circumstances. Fundamentally, this is because of the way that oxygen gains electrons in the process of attaining a more stable, fully "reduced" energetic state—as when it combines with hydrogen to form water. On the way towards this state, extremely reactive intermediates, known as "singlet oxygen", "peroxide" ions, "superoxide" and "hydroxyl" radicals are formed which are liable to react with any chemical species in their vicinity and convert them into free radicals. Free radicals (chemical species with unpaired electrons) react in their turn with other chemicals to produce further free radicals, so causing potentially very complex, autocatalytic, "chain reactions" to occur.

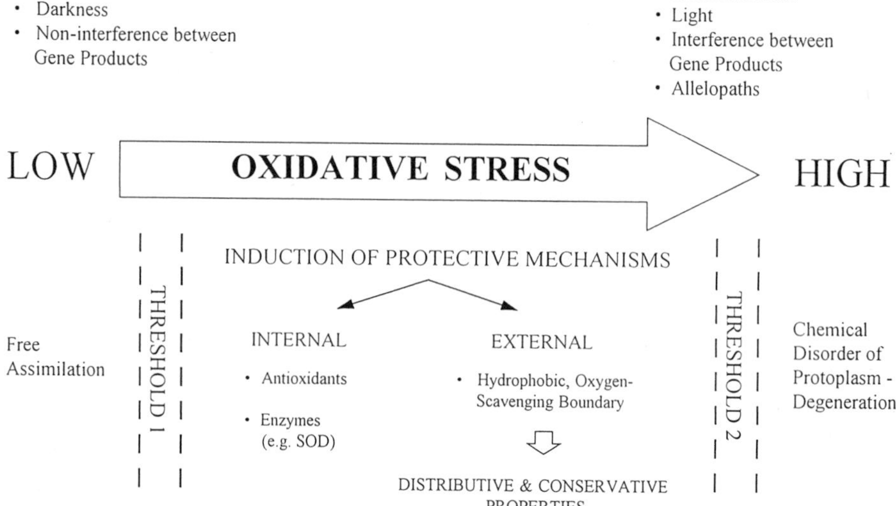

Figure 5.2. The toxic influence of oxygen on boundary properties. Factors contributing to low or high oxidative stress are listed above the arrow. Distinctive responses to different degrees of oxidative stress above and below two thresholds are listed below the arrow. (From Rayner, 1996c).

If oxygen-mediated free radical chain reactions are allowed to run out of control within cells, then the result is a chemical meltdown into a disordered state culminating in death. Cells exposed to oxygen therefore need to circumvent this possibility in some way if they are to stay alive. One way in which they achieve this is by a highly dissipative, energy-consuming, "gas-guzzling" process! Electrons produced by the breakdown of food molecules ("fuel") are passed along a chain of carriers to oxygen during the process of aerobic respiration (see Chapter 2). The chemical energy (ATP) generated during respiration is then used to assimilate more fuel through permeable (and consequently leaky) cell boundaries. Another way is by producing protective "antioxidant" compounds and enzymes like superoxide dismutase (SOD for short) that reduce the reactive intermediates. Finally, the cell can be protected by insulating its boundary in such a way as to lessen access of oxygen to the interior.

If the mechanisms that protect cells from oxygen damage prove inadequate or become dysfunctional, then the cells become subject to "oxidative stress" and on course to degeneration. Any disruption of mitochondrial functioning, for example, can have this effect. On the other hand, incipient oxidative stress may, providing that it does not go too far, provide the cue for inducing the protective mechanisms (Fig. 5.2).

Intriguingly, oxygen may play a part in the boundary-insulating processes that protect cells from its own toxic effects. This is because it is capable of chemically cross-linking or polymerizing hydrophobic compounds produced at cell boundaries, particularly in the presence of phenol-oxidizing enzymes. These enzymes can also cause depolymerization under appropriate conditions of aeration and acidity.

Since insulation also affects the hydrodynamic properties of cell systems, oxygen may be key to the ability of these systems to produce assimilative, explorative, conservative or redistributive states (see also Chapters 4 and 6). In fact, in terrestrial environments, the presence of oxygen in the gaseous phase—through which it diffuses many times more rapidly than through water—may pose a more immediate threat to survival than drying out. To put it another way, the danger of drying out lies not so much in the loss of water as in the increased exposure to gaseous oxygen.

Many features of terrestrial life forms that have been viewed as boundary-sealing mechanisms to prevent water loss could therefore serve more primarily in restricting oxygen access to tissues. These include the production of a thick skin, waxy and corky coverings, and the exudation of gums, latexes and resins. In this sense a tree might well be considered to be a response to oxygen toxicity, as much as a device for connecting sites of water uptake in soil to a distant photosynthesizing canopy!

Similarly, many systems that have been thought to maximize aeration of the interior of terrestrial organisms, such as the lungs of vertebrates and the tracheal system of insects (see chapter 4), might better be viewed in terms of the need to *regulate* oxygen access. This would be consistent with the way they branch from relatively wide openings at the otherwise impermeable external boundary into smaller and smaller branches as they penetrate deeper into the tissues, rather than *vice versa*. In the case of insect trachea, the finest branches are water-filled—and hence do not transmit air to the tissues except at gas-guzzling times of very high metabolic demand.

2. Outside-in Versatility in Determinate Systems

In Chapter 3, a comparison was made between relatively determinate and indeterminate systems. Determinate systems diversify from outside-in, i.e. differentiation occurs within an external boundary that ceases to expand with the passage of time. Indeterminate systems diversify from inside-out, with the external boundary continuing to expand and change its form.

Both kinds of systems can exhibit versatility by means of the reconfiguration and abandonment of boundaries. However, whereas in indeterminate systems new functional states or boundary configurations can remain interconnected, these states become isolated from one another in determinate systems, so that one takes on where the other leaves off.

2.1. Metamorphosis

Perhaps the most striking example of a change of functional state in determinate systems is the conversion of the larval form of an animal into an "adult", e.g. the transformation of a tadpole into a frog, or of a caterpillar into a butterfly. Such transformations broadly correspond with the conversion of assimilative or growing phases into reproductive phases and involve very obvious stages of boundary relocation and reconfiguration.

In the case of a tadpole, the tail and gills which are appropriate to a life in water degenerate and become replaced by the legs and lungs which enable frogs to make their way on land. The degeneration and resorption of the tail is a redistributional process that involves the "prescriptive" or developmentally "programmed" death (also known as "apoptosis") of cells.

Degenerative processes are even more evident during insect metamorphosis, where virtually the entire muscle system of a larval stage such as a caterpillar is absent from the adults. Here, the relocation of boundaries from larval to adult stages involves virtually complete abandonment and redistribution.

Insect metamorphosis also involves a transition from soft-bodied larval forms with relatively deformable external boundaries to hard-bodied forms with a rigidified, armour-like, "exoskeleton". The soft-bodied forms are able to enlarge, partly because of the expandability of their skin or "cuticle" and partly because once the cuticle has reached the limits to which it can be stretched, it is separated off and discarded. Often there are several such moults ("ecdyses") between separate larval stages ("instars"), but each stage is a recapitulation—a larger-scale model—of its predecessor, rather than a fundamentally new form.

As a result of moulting, the growth of insect larvae proceeds by a series of cycles of stops and starts. These cycles are associated with, if not orchestrated by, a counteractive interplay between hormones, related to the hardening off of old cuticle and generation of new cuticle. The time at which the hormones are produced varies with environmental conditions, implying that a potentially complex system-boundary feedback process is involved. One of the first of the hormones to be studied became known as "ecdysone" or "moulting hormone". Originally, it was thought that this hormone alone was responsible for moulting, because it increases in amount during each instar. However, it has subsequently been realized that levels of ecdysone actually decrease at the time when moulting occurs (Fig. 5.3).

When the final larval stage reaches its size limit, the cuticle is hardened by a tanning process involving the action of phenol-oxidizing enzymes, so forming a pupa. The pupa may be thought of as a conservational stage which arrests further expansion and seals in the resources assimilated by the larva. It may be that such a boundary-sealing stage is essential if the reiterative process of larval growth is to be brought to an end, so providing scope for the fundamental reorganization that leads to the emergence of the adult insect. An analogous process, "encystment", is also interposed between distinctive life cycle stages in many animals, including parasites such as flukes which have two or more "alternate hosts". An example of the latter is *Schistosoma*, which alternates between water snails and human beings, causing the widespread tropical disease, bilharzia.

As already indicated, the process of emergence involves both the abandonment of the old boundary—the hardened pupal casing—and redistribution. The latter is associated with degeneration of larval tissues and activation of embryonic cells in so-called "imaginal discs" that have lain dormant during proliferation of larval tissues from the egg.

The degenerative processes involved in metamorphosis have always been thought to be a necessary part of the development of animals with multiple life cycle stages. The idea that these processes are actually "programmed" to occur has therefore been implicitly if not explicitly assumed. In other walks of life, death has usually been regarded as an inescapable and fundamentally undesirable consequence of damage or imperfection, the result of infectious disease, accidents, predation, accumulation of "mutations", toxins, etc.

Recently, however, the idea that programmed cell death is also an important component of the development of animals without multiple life cycle stages has gained favour. Cell death has even been envisaged to serve a "protective" function that limits the indefinite proliferation of unspecialized cells that would otherwise run out of control and cause cancers. Indeed it is their lack of a self-destruct mechanism and their resultant potential immortality that helps to make cancer cells so dangerous.

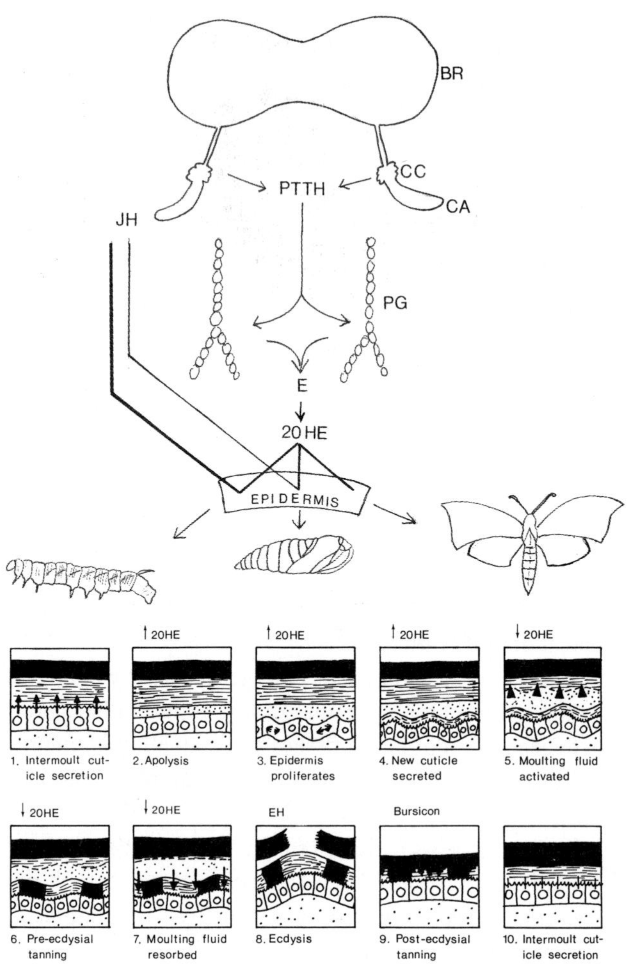

Figure 5.3. Control of moulting in insects. The upper drawings shows how cells in the brain (BR) secrete prothoracicotropic hormone (PTTH) which passes down nerve axons to the corpora cardiaca (CC) which store and later release PTTH. The PTTH stimulates the prothoracic glands (PG) to secrete ecdysone (E), which is converted into the active hormone, 20-hydroxyecdysone (20HE) in the target tissues. Reduction in 20HE levels in combination with large, small or absent amounts of juvenile hormone (JH) produced by the corpora alata (CA) leads to larva-larva, larva-pupa or pupa-adult moults respectively. The lower drawings shows how the sequence of events from apolysis—the separation of epidermal cells from the old cuticle—to final tanning is associated with rising and falling levels of 20HE, and the eclosion (EH) and bursicon hormones. (Based on drawings provided by S.E. Reynolds).

If degenerative or boundary-fixing processes do not ensue during development, an organism may persist in a babyish state, a condition known technically as "neoteny" or "paedomorphosis". A well-known example is an amphibian known as the axolotl, which retains the gills of a tadpole even after it has become reproductive. Even more remarkably, the larval stage of sea squirts or "tunicates" (see Chapter 2) is believed to have given rise, via neoteny, to the evolutionary line from which vertebrates, including human beings, arose. Human beings have in turn sometimes been suggested to be babyish apes, in which the long gap between infancy and "adulthood" allows a relatively long period during which the organisms continue to develop and to learn by experience. The babyishness of domestic animals has also been suggested to underlie their ability to accept rather than reject human beings as co-habitants, masters or mistresses.

2.2. Alternative Phenotypes

Metamorphoses and equivalent phenomena allow an organism to exploit heterogeneity over the course of its life cycle by producing specialized separate stages that serve distinctive functional roles in different kinds of environments. Nonetheless, the versatility of metamorphosing systems is restricted by the fact that one stage can only arise from the other in a specific sequence: e.g. larva -> pupa -> adult -> egg -> larva. Such systems cannot therefore simply switch between modes in whatever unpredictable sequence may be appropriate to actual circumstances. Failure of one mode to locate an environment where it can proliferate automatically implies failure to produce the other mode—however suitable conditions might be for the latter. Selection against one mode therefore implies selection against the other(s). This fact has great practical significance in the control of "pest" organisms, since it is only necessary to target one life cycle stage in order to eliminate the rest. Mosquitoes can be eradicated (at least theoretically) by killing their larvae.

More versatility can be achieved by incorporating a variety of developmental "options" or "alternative phenotypes" into the life cycle. Here, each option may be specialized for a particular function, but there is no set sequence in which one option arises from another. Instead, the selection of options may either be determined by environmental "cues", or options may be selected randomly and those that are most appropriate to the particular circumstances amplified. The latter "error and trial" process may seem more wasteful in that it does not match mode to immediate conditions. However, it is also more "anticipative" in that it allows the requisite modes to be produced in advance of the appropriate circumstances. An example of such an anticipative process has already been encountered in the immune system of vertebrates (Chapter 4). However, the latter system involves irreversible change due to genetic rearrangement in different cell lines within an organism. It does not provide versatility at the "individual organism" level of organization, where changes in phenotype, in being at least potentially reversible, require a change in gene expression rather than genetic content.

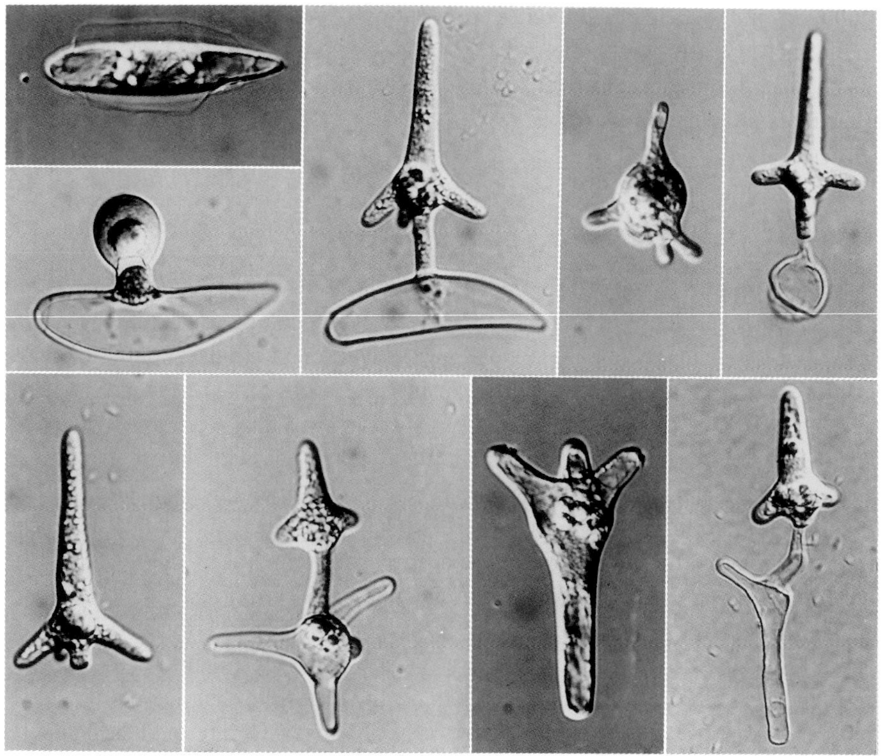

Figure 5.4. Alternative types of spores in the fungus *Erynia conica*. Top row (from left): cornute spore with inflated outer envelope (possibly acting as a buoyancy aid) and producing a globose spore (upper and lower photographs); a cornute spore producing a stellate spore; a globose spore developing projections; a globose spore producing a stellate spore. Bottom row (from left): a stellate spore; a stellate spore producing another stellate spore; a coronate spore; a coronate spore producing a stellate spore. (Courtesy of J. Webster)

A striking example of alternative phenotypes at a relatively simple level of organization is provided by the spores of certain fungi. For example, *Erynia conica*, a parasite of aquatic insect larvae, produces four types of spores (Fig. 5.4). Each type can reproduce either by repetition—producing a spore like itself—or by transforming into another type. Only globose spores can infect insects. Cornute spores are boat-shaped and float on the surface of water, whilst the coronate and stellate types readily become entrapped on water-splashed, moss-covered boulders.

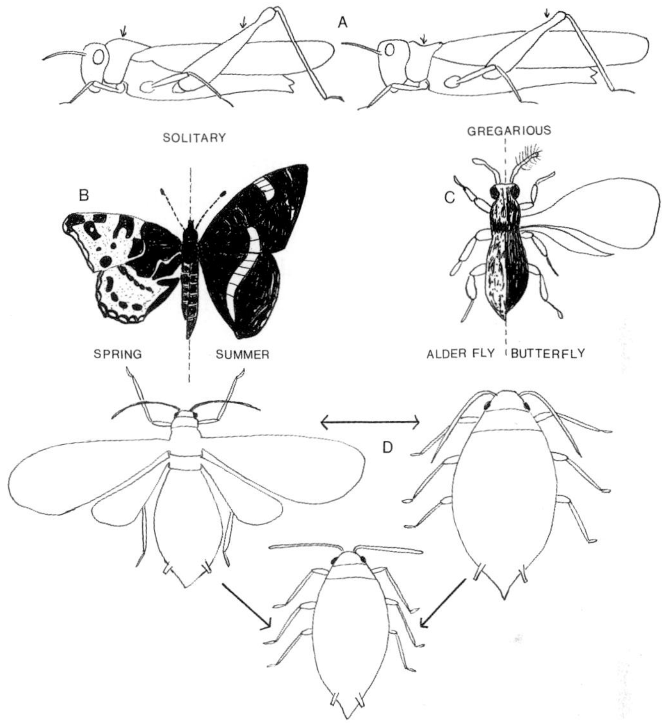

Figure 5.5. An array of alternative phenotypes amongst insects. A, solitary and gregarious females of the migratory locust—arrows show main differences in form, but the greatest phenotypic difference is in behaviour, the one associative the other dissociative. B, spring and summer forms of the Map butterfly, *Araschnia levana*. C, forms of the parasitic wasp, *Trichogramma semblidis*, emerging from alder fly or butterfly hosts (based on Wigglesworth). D, parthenogenetic winged and wingless females (the former produced under crowded conditions) proliferating during long summer days, give rise to sexual forms in the short days of autumn.

There are many other examples of alternative phenotypes (Fig. 5.5). There are protists which develop big or small mouths depending on the size of their food. There are parasitic wasps which do or do not possess wings and bushy antennae depending on which host their larvae grow up in. There are butterflies which have different colours and body patterns depending on the time of year at which they emerge from the pupa. There are the castes and morphs of social insects (Chapter 2;

see also Fig. 6.16), which differ in the way their body boundaries expand depending on how the larvae are fed.

The fact that the relative independence of these alternative phenotypes allows them to be separately exposed to natural selection without affecting one another's chances of proliferation, raises important possibilities. If there is some mechanism by which differences in their boundary properties can be re-inforced or genetically fixed between generations, they may be capable of diverging from one another and so founding separate evolutionary lines. Also, the possibility of producing some alternative life form, whilst retaining the ability to produce one that has already proved fit, provides a way of continuing to innovate without the risk of extinction in the short term. Alternative phenotypes enable developmentally determinate organisms to have their cake and eat it: a basal population can continue to maintain itself at carrying capacity whilst a proportion of its offspring fulfil a role as venturers.

2.3. Evolutionary Remodelling

One of the deeper ruts produced by the application of genetic models to Darwinian theories of evolution by natural selection has been the perception that evolutionary change is brought about solely by alterations in genetic (nucleic acid) content. From this outlook, genes and their products are envisaged to be informational building blocks, analogous to computer bytes. New forms seem to depend on new kinds or new combinations of these building blocks. Evolution is taken to result from computational processes of trial and error in which varied types and arrays of information are cast onto a landscape of box-like niches and those that don't fit rejected. Boundary conditions are preset so that all the system has to do is adapt the "if...then" rules of its genetic software to those conditions in order to survive.

Corresponding with this outlook, mechanisms of mutation, recombination and changes in frequency of genes have been of primary concern in considerations of rates and pathways of evolutionary diversification and adaptation.

Mutation, however, is extremely infrequent (generally of the order of 10^{-9} per nucleotide per cell generation). In the vast majority of cases it is also far more likely to lead to dysfunction than to adaptive benefits.

Recombination, in "shuffling the pack", may allow for some variation on a genetic theme, and hence some versatility at the population level of organization.

However, it is arguable whether recombination in itself can allow both the emergence *and* maintenance of new themes. On the one hand, the homogenizing effect of the removal of "unfit" variants by natural selection in local populations would reduce the scope for complementation needed to allow emergence (recombination between populations from disparate locations or niches may be another matter, see Chapter 7). On the other hand, the disruptive effect of reshuffling on any favourable combinations of genes would prevent maintenance.

Changes in gene frequency brought about by sexual transmission and natural selection may account for the spread or eclipse of individual traits, and resultant small-scale evolutionary shifts. However, on their own, they can do little more than that. The capacity for changes in genetic ingredients in themselves to bring about rapid, large-scale shifts in organizational pattern is limited.

However, there are many indications that radical shifts in organizational pattern can occur. For example, there are many striking examples of mimicry which enable animals to be camouflaged or to resemble something dangerous (see Chapter 7). For such mimicry to be effective it must be brought about rapidly rather than *via* a long line of unadaptive intermediates. The ability of different evolutionary lines to converge on similar shapes, such as those of sharks (fish), ichthyosaurs (reptiles) and dolphins (mammals) similarly implies rapid and co-ordinated changes affecting a large number of characters. Equally, the proliferation of groups such as mammals into a multitude of forms over relatively short geological time scales implies extraordinarily fast rates of evolutionary differentiation.

A clue to the origin of radical evolutionary shifts may lie in the fact that, as already described, there can be distinctive life cycle stages and alternative phenotypes which have identical genetic composition. Also, what might appear to be very different species can have near-identical DNA (human beings and chimpanzees for example). To change form, it therefore doesn't seem to be necessary to change ingredients so much as the context within which these ingredients interact.

The idea that it is not necessary to change information, only to change the way the information is processed, in order to bring about radical changes in output is familiar in many situations. Any cook appreciates that very different products can be obtained from the same set of ingredients—notwithstanding that changing the ingredients (e.g. by leaving one out—simulating the usual consequence of mutation) can also have dramatic (often disastrous) effects! The same is true of genes. Genes define the ingredients, not the rules, governing an organism's pattern. They provide memory, but not necessarily the operating system that processes stored information into viable outputs. The operating system is based on feedback at contextual boundaries. It integrates genetic and environmental inputs in a complex interplay that can bring about radical shifts in boundary configurations without requiring any major, potentially damaging, irreversible changes in information.

There is therefore a good case for understanding evolutionary remodelling in terms of both genetic and organizational processes. The latter may be especially important if there is some means whereby particular sequences of boundary-defining events can be re-iterated. That such means exist is implicit in the occurrence of alternative phenotypes, and perhaps even more so in the developmental versatility of indeterminate living systems.

3. Inside-out Versatility in Indeterminate Systems

3.1. Mode-Switching

Indeterminate organisms are versatile both in the variety of functionally distinctive offshoots—leaves, flowers, polyps, fungal fruit bodies etc—that they can produce, and in the form of the interconnections between these offshoots.

The offshoots are commonly composed of tissues, and are often referred to as "organs" equivalent to those of determinate organisms. However, they are generally produced externally and at varied places and/or times—not internally and once-and-for-all. They therefore correspond more with alternative phenotypes than with organs, although the fact that they remain interconnected makes it easy to view them as components of the same system.

In fungi, determinate offshoots from the indeterminate mycelium generally consist of specialized absorptive, storage and reproductive structures. The absorptive and storage structures are often either bulb-shaped or consist of progressively finer branches: good examples of both types are provided by the fungi which form "vesicular-arbuscular" mycorrhizas (Fig. 5.6; see also Chapter 7). The reproductive structures are spore-producing bodies or "sporophores". They may be anything from single compartments of hyphae to elaborate proliferations that are readily discernible to the naked eye as "fruit bodies"—toadstools and the like.

Often, the same fungus can, depending on circumstances, produce very different kinds of sporophores—a fact which is both an expression of versatility on the part of the fungus and a source of confusion to human beings! Sporophores, as the outward signs of a hidden infrastructure (the mycelium), are generally used by people both to find and to identify fungi. Therefore, when different kinds of sporophores are found, it has been normal to assume that they belong to different fungi, and therefore to assign different names to them. Only by observing the same fungus growing under a range of conditions has it been possible to clarify the connection between its different reproductive states—by which time the fungus has usually come to be known by at least two, and often more, names! Such are the trials of determinate organisms (human beings) trying to pigeon-hole indeterminate ones!

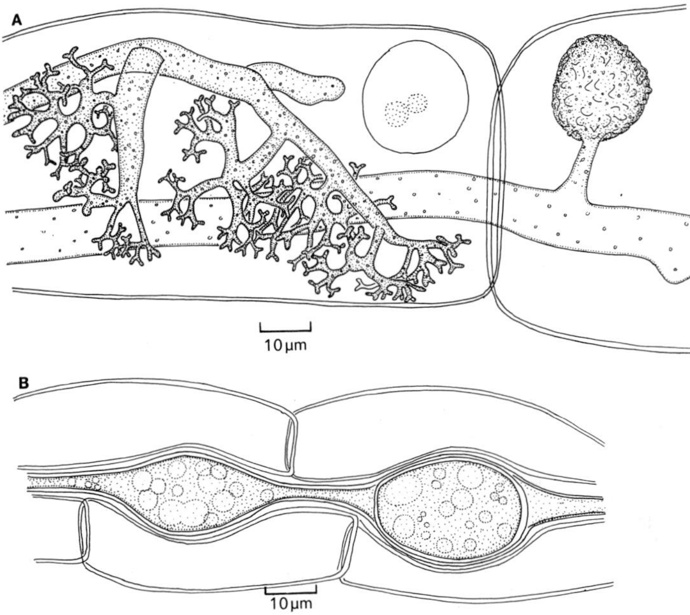

Figure 5.6. The structure of vesicular-arbuscular mycorrhiza. A, onion root cells infected by *Glomus mosseae*—the cell to the left contains a nucleus with two nucleoli and a branched haustorium or arbuscule, and the cell to the right contains a degenerating arbuscule; B, vesicles from roots of *Arum*. (From Webster, 1980)

Many fungi produce both asexual (mitotic) and sexual (meiotic) reproductive states (Fig. 5.7). By so doing, they achieve three kinds of versatility. Firstly, they can both maintain and change genetic identity in the course of proliferating (cf. Chapter 4). Secondly, they can adjust the scale on which they invest in reproduction, generally because the asexual structures are smaller and more diffusely distributed than the sexual ones. Thirdly, the kinds of spore produced can be varied (generally in their size and degree of insulation) so as to equip them for different roles in the short- and long-range dispersal of the organism through space and time.

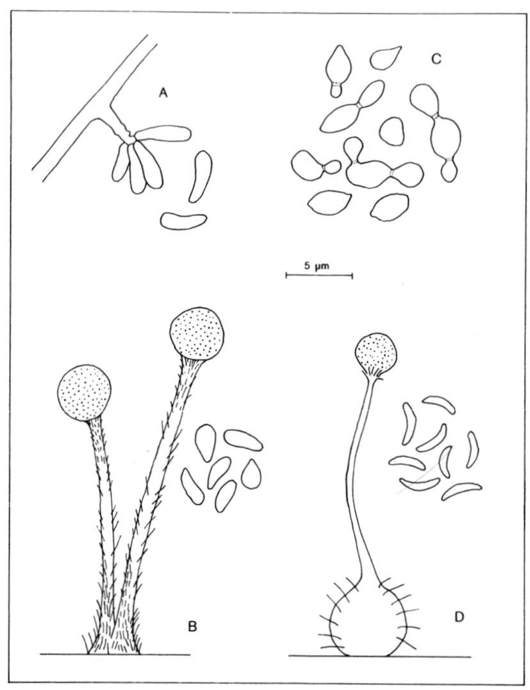

Figure 5.7. Sexual and asexual spore-producing states of the Dutch elm disease fungus, *Ophiostoma novo-ulmi*. A, asexual "*Sporothrix*" stage in which bunches of pear-shaped spores ("conidia") are generated from individual hyphae. B, asexual "synnemata" (c. 1 mm tall) consisting of individual stalks, each consisting of many parallel-aligned hyphae and bearing a droplet containing pear-shaped conidia. C, asexual, budding yeast-like stage which proliferates within the sap stream of infected trees. D, sexual stage consisting of a flask-shaped "perithecium" (c. 0.5 mm tall) from which banana-shaped, meiotically produced spores ("ascospores") ooze out at the top of a long, neck-like tube. (Drawn by C.M. Brasier, from R.J. Stipes and R.J. Campana, 1981.)

The possession of several different spore-producing stages is characteristic of many disease-causing fungi. The devastating effect that the fungi causing Dutch elm disease have had on the landscape is at least partly related to the fact that they produce four different kinds of sporophores (Figs 5.7, 5.8). One type is responsible for distributing the organism within the sap stream of infected elm trees. Two other types have a form which eases the pick-up and carriage of spores from stalk-like structures by the bark beetles that transfer the fungus from infected to uninfected trees. One of these types generates spores mitotically, whereas the other generates spores meiotically.

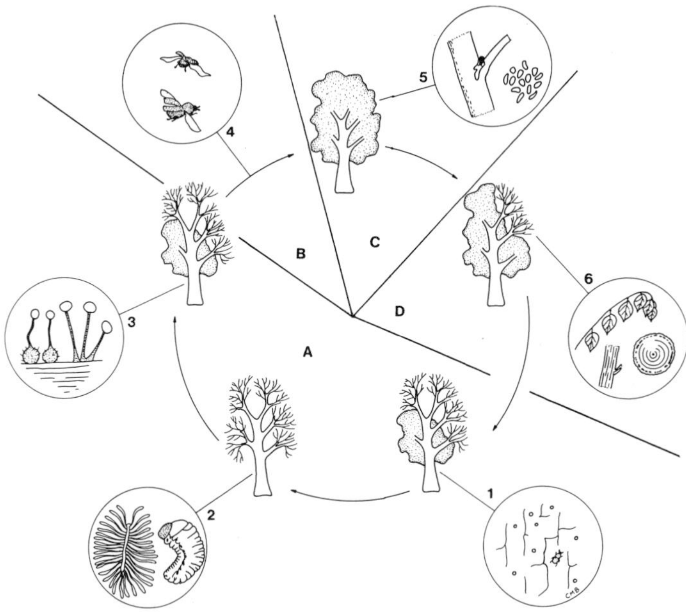

Figure 5.8. The main growth and dispersal phases of the ascomycete fungus, *Ophiostoma novo-ulmi*, in relation to the annual cycle of Dutch elm disease. A, *Bark phase*. In late summer and autumn, trees weakened by the disease become breeding sites for bark beetles (1). The beetle larvae excavate galleries in the bark (2) and *O. novo-ulmi* fruits in the breeding galleries (3). B, *Flight phase*. In spring and summer, adult beetles emerge carrying spores of *O. novo-ulmi* (4). C, *Feeding groove phase*. The newly emerged beetles fly to feed in twig crotches of healthy elms (5) and by so doing can introduce the fungus into the sapstream. D, *Wood phase*. Infected twigs and branches wilt, showing characteristic streaks or spots in the infected annual ring (6). (From Brasier, 1986)

An extreme case of multiple spore-types is provided by the so-called "rust fungi", which can produce up to five different kinds of spore-producing stages—pycnia, aecia, uredia, telia and basidia—often on two different kinds of host plant ("alternative hosts"). These fungi therefore parallel animals with multiple life cycle stages, especially certain animal parasites, such as flukes which produce different life cycle stages in different hosts, e.g. *Schistosoma* (see above).

An example of a rust fungus is "black stem rust", *Puccinia graminis*, which alternates between grasses (including cereal crops) and barberry plants. The basidiospores, which transfer the fungus from grasses to barberry, are produced meiotically, whereas the remaining spore types, including the aeciospores which transfer from barberry to grasses, are produced mitotically.

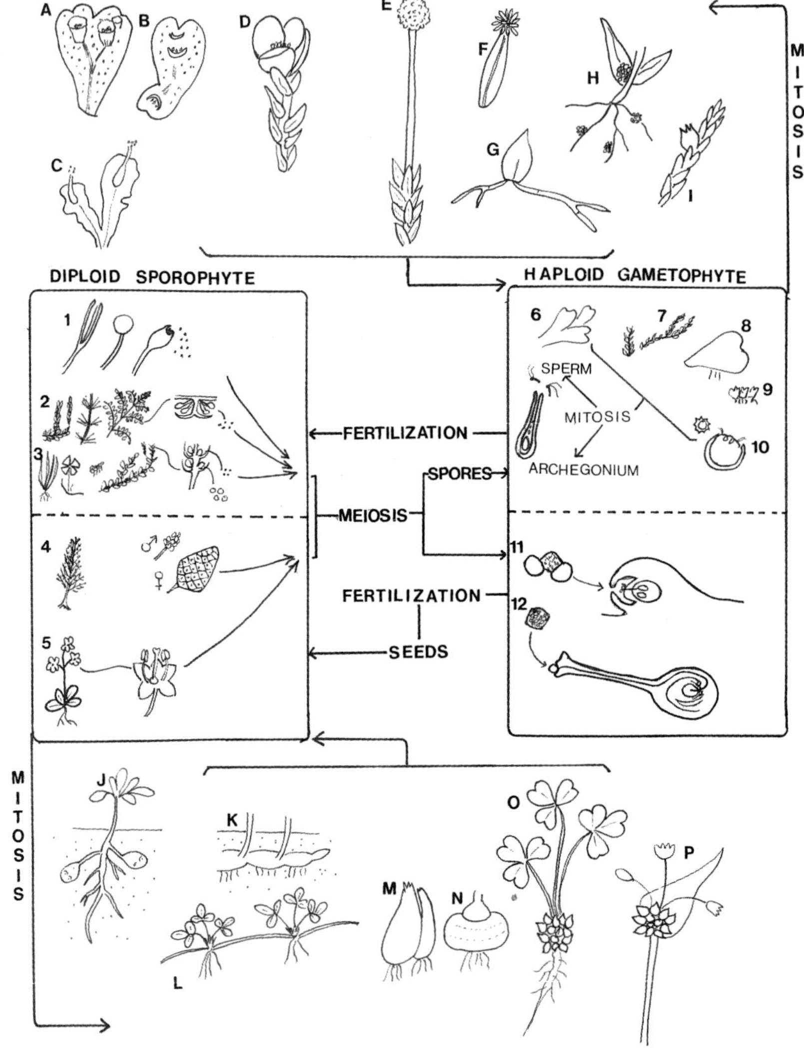

Figure 5.9. Meiotic (sexual) and mitotic (asexual) reproduction in land plants. Symbols for meiotic pathways: 1, spore capsules of (left to right) hornworts, liverworts and mosses; 2, shoots and fronds of (left to right) club mosses, horsetails and ferns—and a section through a "sorus" of male fern containing a group of sporangia protected by an "indusium"; 3, shoots and fronds of quillworts, water ferns and *Selaginella*—and a section through a cone of the latter containing micro- and megasporangia; 4, a gymnosperm tree, such as pine, with

male and female cones; 5, an angiosperm with flowers; 6, "thalloid" vegetative form of hornworts and some liverworts; 7, leafy vegetative forms of mosses and some liverworts; 8, "prothallus" of ferns; 9, "tuber" of club mosses; 10, mega- and microspores of *Selaginella*; 11, pollen grain and megasporangium (ovule) of a gymnosperm, such as pine; 12, pollen grain and stigma, style and megasporangium (ovule) of an angiosperm. Symbols for mitotic pathways: A,B,C, liverwort thalli of *Marchantia*, *Lunularia* and *Blasia* producing "gemmae" in cup-, crescent- and flask-shaped containers; D, gemma cup of the moss, *Tetraphis pellucida*, consisting of a rosette of large leaves at the top of the stem; E, drumstick-like gemma head of *Aulacomnium androgynum*; F, leaf-tip gemmae of the moss, *Ulota phyllantha*; G, detached leaf of *Funaria* moss, sprouting filamentous protonemata; H, gemmae of the moss, *Bryum rubens*, in leaf axil and on protonema (whence known as "tubers"); I, "bulbil" formed in leaf axil of the moss, *Pohlia annotina*; J, potato plant with underground "stem tubers"; K, underground rhizome of Solomon's Seal; L, strawberry plantlets interconnected by stolons; M, bulb and daughter bulb of daffodil; N, corm of crocus; O, bulbils produced at the base of wood sorrel; bulbils produced at the top of a flowering stem (scape) of garlic.

This kind of life cycle, in which there is a full complement of spore types and two alternate hosts is known as "macrocyclic". Another example of a macrocyclic species is *Cronartium ribicola*, which alternates between currant bushes and five-needled pines. This fungus has caused devastating epidemics of "white pine blister rust" in North America during the twentieth century. One way of attempting to control diseases caused by macrocyclic rusts is by eradicating the alternate host—i.e. currant bushes in the case of white pine blister rust. However, this is not always very practical and can be both economically and ecologically costly.

Other kinds of rust fungi are known as "microcyclic", in that they lack one or more of the spore stages in the life cycle and infect only one kind of host. They appear therefore to have been able to "short-circuit" their life cycle, but in so doing may have lost some of their versatility.

Plants, like fungi, can reproduce both meiotically and mitotically (Fig. 5.9). Meiosis either leads directly to the production of male and female gametes (as in some algae) or to the production of spores within structures known as "sporangia". Mitotic reproduction, on the other hand, involves producing groups of cells or "plantlets" that can mature into adult plants. These plantlets can arise from leaves or stolons, or from various kinds of "storage organs" (rhizomes, tubers, corms and bulbs). This kind of asexual reproduction is often called vegetative reproduction.

In the "bryophytes" (mosses and liverworts), the meiotically-produced spores are all of the same type ("homosporous") and germinate to form a haploid plant or "gametophyte" which grows independently and eventually produces male and female sex organs ("antheridia" and "archegonia" respectively). Fertilization of the egg in the female sex organ results in the outgrowth, virtually as a parasite on the gametophyte, of a diploid "sporophyte" from which further spores are produced.

Many "pteridophytes" (ferns and allies) also produce homogeneous spores which germinate to produce a small, independently growing, haploid gametophyte or "prothallus". However, unlike bryophytes, the sporophyte which emerges from fertilized archegonia on the prothallus, soon becomes an independent plant in its own right, producing the fronds and rhizomes which are characteristic of these organisms.

In some other pteridophytes, however, the spores are of two types ("heterosporous")—small "microspores", which produce male gametes, and large "megaspores", which produce archegonia without forming an independently growing gametophyte. This trend for reduction of the gametophyte stage is continued in the seed plants, the "gymnosperms" (conifers and allies) and "angiosperms" (flowering plants). Here the microspores are known as "pollen" and, typically, fertilization of egg cells produced by the megaspores gives rise to seeds. In some cases, known as "apomicts", however, seeds can be formed without fertilization and meiosis.

Determinate outgrowths from the main stems of plants are generally referred to either as "leaves" or as "trichomes". The latter include various kinds of hairs and thorns. Leaves can assume a wide variety of forms and functions both in the same and in different plants. It is also important to appreciate that what may look like leaves on different plants are sometimes analogous structures, derived by convergent evolution, rather than homologous structures produced by basically the same developmental processes. The "leaves" of seaweeds and mosses, for example, are analogous to rather than homologous with those of vascular plants (pteridophytes, gymnosperms and angiosperms). Also, some structures that look like leaves, such as the "cladodes" of Butcher's Broom, are in fact flattened stems.

In vascular plants, the variety of leaf form and function amply demonstrates the importance of being able to produce alternative phenotypes without changing genetic information. The structures that people most readily think of as leaves are flat or needle-shaped (as in conifers) green structures whose primary function is photosynthesis. Even here there is tremendous variation in shape (Fig. 5.10). Leaves may be lobed or unlobed. They may be simple, consisting of a single blade or "lamina", or compound, dissected into self-similar series of leaflets. They may differ, even on the same plant, according to the circumstances under which they have developed—a condition known as "heterophylly". For example, leaves developed in strong sunlight ("sun leaves") tend to be thicker but narrower and held in a less horizontal orientation than those produced in shade ("shade leaves"). In ivy (*Hedera helix*), the leaves on flowering stems are unlobed whereas the leaves on non-flowering stems are lobed. Many aquatic plants produce distinctive kinds of leaves below, at and above the surface of the water.

Figure 5.10. Some of the variations in shape produced by the foliage leaves of plants. A, pinnate compound leaf of ash (*Fraxinus*); B, palmate compound leaf of horse chestnut (*Aesculus*); four-lobed leaf of tulip tree (*Liriodendron*); dissected bi-pinnatifid leaf of *Anemone*; E, cordate (heart-shaped) leaf of celandine (*Ranunculus ficaria*); F, runcinate leaf of dandelion (*Taraxacum*); peltate leaf of marsh pennywort (*Hydrocotyle*); H, water crowfoot (*Ranunculus aquatilis*) with lobed emergent leaves and dissected underwater leaves; I, arrowhead (*Sagittaria*) with arrow-shaped aerial leaves, oval floating leaves and strap-shaped underwater leaves. (Not to scale)

Figure 5.11. Some examples of leaf modifications. A, section through a daffodil bulb showing outermost scale leaves, inner storage leaves, innermost foliage leaves, and buds; B, part of a stem of *Berberis*, showing groups of spine-tipped foliage leaves with three leaf spines at their bases; C, part of an *Acacia* stem with pinnate foliage leaves and "phyllodes" (flattened leaf stalks); D, pinnate pea leaf with large "stipules" at its base and coiled "tendrils" in place of leaflets at its tip; E, a flowering stem of wood spurge, with elongated foliage leaves, bracts—which resemble petals—and very small flowers consisting of single anthers and ovaries; F, rosette of butterwort—the leaves are sticky and have inrolled margins; G, leaves of Venus fly-trap, with open and closed traps; H, sundew—leaves with glandular hairs; I, bladder of bladderwort, with trapped water flea; J, pitcher of *Nepenthes*—formed on a tendril produced at the end of a foliage leaf. (Not to scale)

Structures that can be recognized to originate developmentally as leaves can also undergo a variety of reconfigurations which suit them for distinctive functional roles (Fig. 5.11). The various parts of flowers, i.e. the sepals, petals, stamens and carpels can all be thought of as modified leaves. Leaves produced at the base of flowering stems are known as "bracts", and can sometimes themselves become brightly coloured, taking over the role of petals in the attraction of pollinators. Leaves, and parts of leaves can be modified into the coiling tendrils of some climbing plants. They can also be modified into protective scales around buds, into storage organs as in bulbs and even into spines which deter herbivores. Some of the most extraordinary leaf-modifications occur in plants which extend their supplies of nitrogen by capturing and digesting insects and crustacea—butterworts, bladderworts, sundews, venus fly traps and pitcher plants.

The indeterminate axes from which the determinate offshoots of plants and fungi arise can themselves exhibit a remarkably versatile range of alternative forms or states. Transitions between these states during development can occur either gradually or abruptly (whence they appear as "switches") and may generally be regulated by changes in boundary permeability, deformability and internal partitioning. These changes may in turn be brought about by oxygen-mediated polymerization and depolymerization processes at cell boundaries. Once these processes have been initiated, they become autocatalytic, due to the generation of free radicals, and therefore effectively uncoupled from genetic control.

In fungi, a variety of mode switches can be identified (Chapter 4; Fig. 4.12), and in some cases all may be expressed by the same organism at some time or place during its development. The fungus may proliferate as cells or as mycelium. The mycelium may be densely branched, but slowly extending or effusely branched and more rapidly extending. The hyphae may be septate or non-septate, anastomosed or non-anastomosed, assimilative or non-assimilative, diffuse or aggregated.

Mode switches in the indeterminate root and shoot systems of plants are less easy to itemize than in fungi. However, abrupt changes in branching pattern from the equivalent of slow-dense to fast-effuse forms are commonly observed, for example at the onset of flowering or during the outgrowth of stolons. Plants with predominantly slow-dense patterns have sometimes been described as "phalanx strategists" and those with fast-effuse forms as "guerilla strategists". Plant root systems are often divided into relatively highly branched, absorptive "short roots" of limited duration and less branched, indefinitely extending, conducting "long roots". Equivalent alternations occur between stoloniferous and rooting stages of whole plants such as strawberries and can be likened to those between the nomadic and settled stages of animal societies, such as those of army ants (see Chapter 3). The arrays of flowering stems ("inflorescences") produced by many plants often have characteristic branching patterns ("umbels", "corymbs", "racemes" etc), and indeed differences between these patterns are often used in classification. However, the patterns can sometimes be transformed readily from one to another experimentally, for example by treatment with plant hormones, showing that they are susceptible to remodelling.

All the examples of mode-switching mentioned so far relate to developmentally or socially indeterminate systems. It is therefore interesting to question whether equivalent processes could apply to other kinds of indeterminacy, for example phylogenetic and mental indeterminacy (Chapter 3). In phylogenetic trees, the determinate offshoots could represent highly specialized organisms that are unable to evolve further having reached the limits of adaptive refinement. Alternations between rapid advancement after introduction of new themes, and diversification or proliferation of variations upon these themes would be equivalent to those between fast-effuse and slow-dense phases. Similarly, mental processes often seem to alternate between explorative leaps and consolidatory analysis, with individual ideas perhaps being equivalent to determinate offshoots.

3.2. Autodegeneracy

However versatile they may be, indeterminate systems can still get into a rut as a result of reiterative processes which cause established, retentive structures to develop that increasingly impede and may ultimately arrest further expansion.

It is therefore understandable that a wide variety of autodegenerative mechanisms occur in indeterminate systems. These mechanisms not only enable redistribution to occur, but can also provide the pathways through which redistribution takes place.

Autodegenerative processes in fungi have already been mentioned in Chapters 3 and 4 (see Figs. 3.2 and 3.3). Here they allow a mycelium to forage efficiently amidst arrays of nutrient-rich and nutrient-poor sites, and to maintain the expansion of fairy rings.

In many plants, the action of phenol-oxidizing enzymes can insulate cell boundaries with lignin. Combined with the degeneration of protoplasm this leads to the production of that familiar, predominantly dead, but fundamentally important redistributive tissue known as wood. Plant stems and roots therefore degenerate from inside-out, with the formation of wood keeping pace with outward expansion due to secondary thickening (see Chapter 4). In mature trees, the redistributive process is continued through the agency of fungi which harness the destructive power of oxygen to decay the core wood, causing "heartrot". Heartrot results in the hollowing of tree trunks which provides a huge variety of habitats for animals as well as allowing the tree to recycle itself by proliferating roots within its own internal "compost" of decomposing remains that accumulate within the cavity.

Degeneration also often occurs amongst the central branches of plants such as heather so that an annulus analogous to that of fairy rings is formed. Similarly, structures such as stolons often degenerate as the plantlets that they once interconnected become established in their own right.

Plants also degenerate on the outside, where leaves and branches are drained of their resources, senesce, die and ultimately become detached as they get left behind the inexorably expanding boundary of the system. The final act before detachment usually involves the formation of a corky sealing-off or "abscission zone".

The enormous scale on which these losses are sustained is easily overlooked by eyes focused solely in the present. However, they may be envisaged when viewing a mature tree by imagining the huge number of branches that must have been produced during the course of its lifetime. Only a relatively few of these branches persist and become reinforced into main thoroughfares, the trunks and high order branches. If it were not for continuous self-pruning (often aided by external factors such as storms, decay fungi and arboriculturalists) all trees would be dense thickets.

Phylogenetic trees, likewise, would presumably be dense thickets if it were not for extinctions. However, this is not to say that extinctions should be regarded complacently. Whilst some loss may be vital to further development, there is in any finely balanced system always the danger that degenerative processes may overtake expansive ones and lead to total demise.

The same might be said about the degenerative process which leads to forgetting; it seems likely that some loss of memory is vital to versatile thinking—but total amnesia means, literally, oblivion.

3.3. Opportunism and Episodic Selection

By virtue of their inside-out versatility, indeterminate sytems are able to capitalize on opportunities—the chance to produce the appropriate structural, functional or behavioural form to suit any change in circumstance.

Some of these opportunities occur sufficiently recurrently to be predictable happenings—even if the exact time and place of their occurrence is uncertain. They pose what may be described as "routine selection", and can be anticipated by pre-programming. Such pre-programming forms the basis for a set of ecological strategies in which systems are genetically predisposed to exhibit particular kinds or sequences of patterns. These will be discussed in Chapter 6.

However, there is another kind of change in circumstances which is so novel and/or irregular in occurrence as to seem to "come out of the blue". The only thing that is predictable about such events is that they will happen; when, where and in what form they will take place is utterly uncertain. Such changes in circumstance pose what can be called "episodic selection" and it is indeterminate, or, at least, easily remodellable forms which are best placed to exploit the opportunity that this form of selection provides.

An excellent example, and indeed the one that caused the term "episodic selection" to be coined, is provided by the Dutch Elm Disease fungi. Once considered to be a single species, it has become clear that these fungi are subdivided into three distinctive populations. One of these, *Ophiostoma ulmi*, is relatively non-aggressive, is genetically heterogeneous due to regular sexual outcrossing (see Chapter 4) and in recent decades has been in balance with its host. By contrast, two other populations, one of Eurasian (EAN) and the other of North American (NAN) origin are highly pathogenic (disease-causing) and are responsible for current global epidemics (pandemics). These populations have recently been recognised to be members of a new species, *Ophiostoma novo-ulmi*. The epidemic fronts of both EAN and NAN are genetically homogeneous (clonal or near clonal) whereas populations behind the epidemic fronts are genetically heterogeneous.

The exact origin(s) of *O. novo-ulmi* is (are) uncertain, but seem likely to have involved the transfer, by human beings, of infected material across continental barriers. Such transfer has the effect of introducing the fungus to a previously unexposed—and therefore uniformly susceptible—host population. The resulting selective vacuum would provide just the right kind of opportunity for a new variant to emerge and spread, aided by a switchover to predominantly clonal modes of proliferation.

CHAPTER 6

BALANCE AND CIRCUMSTANCE

1. Four Fundamental Processes—Conversion, Regeneration, Distribution and Recycling

A fundamental inference from the previous two chapters is that through the interplay of differentiation, integration and degeneracy, all living systems undergo cyclic patterns of change. These patterns are brought about by four fundamental processes that vary the flow of energy within and across contextual boundaries in different but complementary ways: *conversion*, *regeneration*, *distribution* and *recycling* (Fig. 6.1).

Conversion processes characteristically follow a phase during which energy has been gathered in ("assimilated") from the external environment. They both immobilize and seal the boundary of a system, or segment of a system, so conserving energy and producing "dormant" survival capsules. Seeds, spores, cysts and storage organs of various kinds all result from conversion processes. However, the degree to which these structures are sealed off from the environment—and they can never be absolutely sealed—varies greatly in different organisms and in different circumstances.

Distribution involves the sealing but not the immobilization of parts of the boundary of a system which are connected, directly or indirectly, to assimilative regions. Resources taken in through the assimilative regions can then drive expansion of the sealed components which are thereby able to negotiate restrictive environments that would not otherwise sustain growth. Explorative structures of all kinds are driven in this way. Examples include the mycelial cords and rhizomorphs of fungi, the stolons of plants and hydroids, the foraging parties of ants and, in human societies, the advance guard of armies and founders of movements.

Regeneration allows the resumption of energy-gathering processes, through the production of open, mobile boundaries, when supplies of available resources are either renewed outside a survival structure or encountered by an explorative or dispersal structure. Examples are provided by all kinds of spore and seed germination as well as the establishment of settlements by migratory systems.

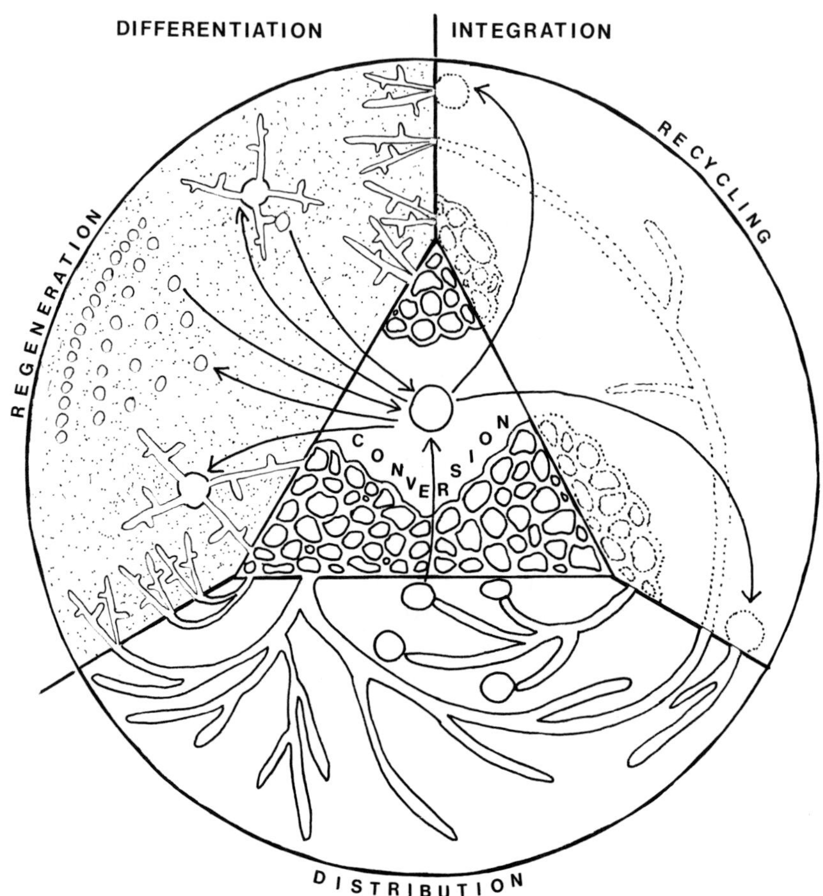

Figure 6.1. Four fundamental processes that regulate energy flow through living systems, and the relation of these processes to resource availability and the properties of contextual boundaries. The three dynamic processes of *regeneration*, *recycling* and *distribution* are depicted as separate domains around a central triangular domain representing the fourth process of *conversion* into dormant survival units and persistent networks. Stippling represents available external resource supplies which promote regeneration and proliferation either as tributary-like branching systems or determinate units. Thin lines indicate permeable contextual boundaries, thick lines impermeable boundaries and dotted lines degenerating boundaries.

Recycling occurs when internal partitioning allows resources to be redistributed from locations that no longer participate in energy-gathering or exploration to sites where these processes are being sustained. It is therefore associated with the degeneration of former boundaries, such as occurs during metamorphosis, fairy ring-formation and amnesia (Chapter 5).

Feedback mechanisms determine which of these four fundamental processes operates at a particular place and time at the boundary of a system. These mechanisms generally lead to boundary-opening when external resource supplies are readily available and various forms of boundary-closure or "self-integration" (see below) when supplies are restricted. The exact nature of these mechanisms differs greatly between and even within systems. Many may be based on responses to supplies of oxygen and reducing power, as described in chapter 5. Others may be due to the direct physical interaction between a dynamic system and its environment that results in the formation and reinforcement of paths of least resistance described in chapter 3. Although these diverse mechanisms produce an enormous variety of complex patterns of structure, function and behaviour, it is important to remember that the processes underlying these patterns remain very simple and common to all systems.

The occurrence of these four processes and the boundary-opening and boundary-closing mechanisms associated with them is vital to the continuation of life in environments where resource availability varies in space and time. It underpins both the production of different life cycle stages by determinate organisms and the versatility of indeterminate life forms discussed in chapter 5.

Nonetheless, the vast majority of ecological and evolutionary thinking during the twentieth century has focused on only one of these processes, namely regeneration. This imbalance has arisen from the neglect of context that comes with taking energy supply for granted and leads inexorably to the gene-centred view that all there is to life is proliferation. When context is ignored, the heterogeneity so evident in real environments is regarded as the result of random imperfection that can either be averaged out by various statistical methods or eliminated entirely by working in suitably "purified" artificial conditions. The world can then be treated ideally as an uncluttered stage on which organisms enact their lives as fully discrete, self-replicating units. All is then fully predictable and calculable. Success is taken to imply good calculation, due to the action of good genes. Failure is regarded as the consequence of bad calculation.

Even supposedly holistic concepts have conformed to this kind of thinking: recent examples include "self-organization" and "complexity" theories, both of which imply that order can emerge from chaos through the interplay of large numbers of simple units. In self-organization theory, this order results from the formation of "dissipative structures" that maximize energy loss from open systems sustained far from thermodynamic equilibrium by very high rates of energy input.

These theories assume high rates of input to organizations composed of fundamentally discrete units, and as such describe patterns relevant only to regenerative processes. They do not recognize the role played by dynamic boundaries *both* in the *emergence* and the *sustainment* of ordered structures. Indeed, they imply that, due to dissipation, life cannot persist for any length of time in resource-restricted environments.

However, processes of boundary-fusion, boundary-sealing and boundary-redistribution all provide means for reducing dissipation, allowing energy to be maintained within the system rather than lost to the outside. Since they lead to more coherent, more persistent organizations in which the discreteness of individual units is blurred, these processes may be referred to as "self-integrational".

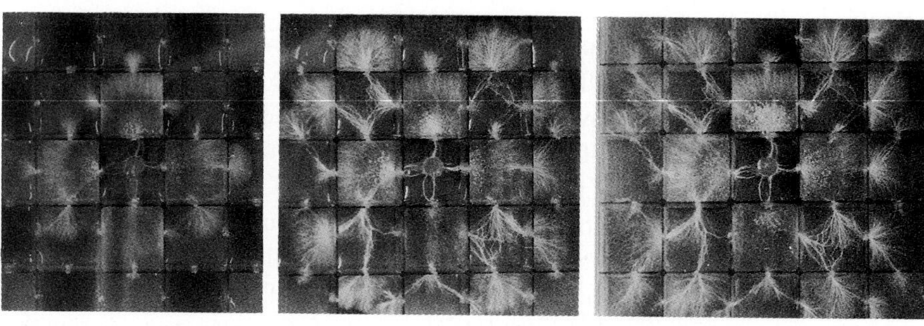

Figure 6.2. Stages in pattern-formation by mycelium of the magpie fungus, *Coprinus picaceus*, when grown through a matrix of alternating high and low nutrient-containing chambers. The matrix is of the same kind as shown in Fig. 3.2, except that the central chamber contains a low nutrient medium.

Variations in the balance between regenerative and self-integrational processes during the exploration and exploitation of heterogeneous environments are epitomized by the mycelial patterns produced by some kinds of fungi, first introduced in Chapter 3 (Figs. 3.2, 3.3). Figs. 6.2-6.4 show the formation of mycelial networks by different fungi when grown in heterogeneous matrices of nutrient-rich and nutrient-poor sites. The same basic themes of dense proliferation on nutrient-rich sites and distribution, redistribution and establishment of persistent networks across nutrient-poor sites are followed in all cases. However, the balance between these themes varies in each fungus, resulting in characteristically different patterns. This suggests that each fungus is primed by the limiting conditions specified in its genes to react to environmental circumstances in a unique way.

The pre-setting of the balance between the four processes in such a way as to cause living systems to produce individually distinctive patterns of response to the same set of environmental circumstances is mirrored in certain manufacturing industries. For example, no two makes of cars are built for exactly the same road conditions, even though their design is based on similar dynamic principles. The capacity for variation upon a theme is as fundamental to the manufacture of a product range as it is in the generation of biological diversity.

Figure 6.3. Two examples of development of mycelium of the basidiomycete fungus, *Coprinus radians*, when grown through a matrix. Notice the formation of fruit bodies and fruit body initials in low nutrient chambers. (Photograph by Timothy Jones)

So, how can—and why should—organisms be primed in these different ways?

2. Differing Emphases—Ecological Strategies

The resources which furnish the growth needs of living things occur in a wide variety of environmental settings; correspondingly diverse patterns of organization are needed to negotiate the conditions within each of these settings to best effect. Just as the different requirements of urban and long-distance driving determine car design, so the need to relate organizational pattern to environmental conditions is key to the range of form of organisms in natural habitats. Also, whereas organisms, like cars, require the versatility to perform under a variety of conditions, their development or behaviour may be biased so as to adapt to those conditions most likely to be encountered. Range Rovers can be driven in cities, but waste fuel and are difficult to park because their overall design is biased towards the needs of traversing rough terrain. Superminis on the other hand are more effective in cities but will fall apart if driven for more than a few miles over rough terrain—unless they have been suitably modified for rally driving! Similarly, the different emphases on assimilation, conservation, exploration and redistribution exhibited by fungi in matrices (Figs 6.2 to 6.4) may reflect adaptive attunement to the kinds of circumstances that they typically encounter in their natural niches.

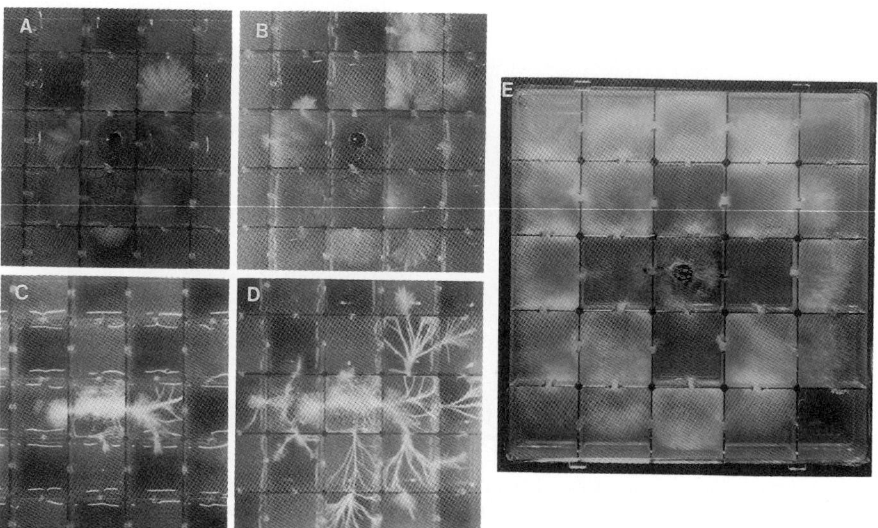

Figure 6.4. Mycelia of various fungi grown in matrices. A and B, C and D; early and later stages of development of the basidiomycetes *Trechispora vaga* and *Phallus impudicus* respectively showing highly heterogeneous, asymmetric pattern and strong commitment to explorative mycelial cord formation (courtesy of Louise Owen). E, the ascomycete, *Daldinia concentrica*, showing sparse sparse development on the first series of low nutrient chambers, but dense development (implying redistribution) in the second series of low nutrient chambers. (Courtesy of Z.R. Watkins)

The environmental determinants which are generally considered to affect patterns of organization most critically are of three main kinds: "disturbance"; "stress" or "adversity", and the presence of competitors (Fig. 6.5). Like the three primary colours, these determinants can be mixed in varying proportions to correspond to a potentially infinite variety of niches. In working out how pattern is related to niche, the art is therefore to identify those sets of attributes which are most apt when each of the three primary determinants operates alone. Each of these sets of attributes can be related to boundary properties and ascribed to what has been called a "primary ecological strategy" (Fig. 6.5). However, I must point out that the use of the word "strategy" here should not be taken to imply any kind of "conscious" forward planning.

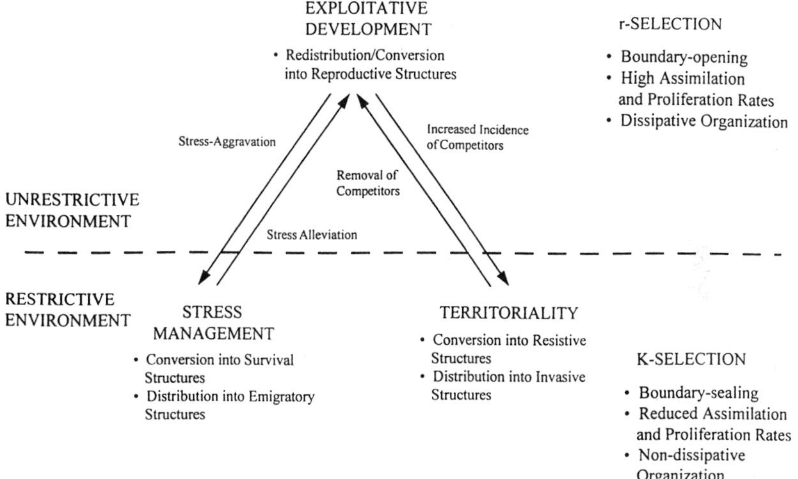

Figure 6.5. Relationships between boundary properties and colonization strategies in restrictive and unrestrictive environments (From Rayner, 1996c; cf. Fig. 6.1).

Ecological strategy concepts provide a very useful way of contrasting the roles of different organisms (or parts of an organism) within a common arena. However, they should be used relatively and not as a means of absolute classification. For example, fungi inhabiting wood can be compared with one another with respect to the degree to which they are favoured by disturbance and stress, so enabling their relative position in "successions" (see below) to be understood. By contrast, it is pointless to compare dolphins with elephants because each of these kinds of animals would be equally "stressed" in one another's habitats.

2.1. *Weeds*

Disturbance can be thought of as any more or less sudden environmental circumstance which makes resources newly available for exploitation. It is often caused by human beings. Fire, toxic pollutants, extreme weather conditions, wave action and ploughing all provide examples of "destructive disturbance" which eliminate established biological communities or parts of communities. By contrast, taking the lid off a pot of jam, pruning a tree, and burying animal or plant remains in soil are examples of "enrichment disturbance" which adds or exposes resources to living systems.

Disturbance therefore results in temporary plenty. In so doing it causes what is known as "R"- or "*r*"-selection (the latter referring to the "*r*" term of the logistic equation, see Chapter 4). This kind of selection favours open-bounded entities that are quick to arrive on the scene, exploit the most readily assimilable resources and then reproduce as supplies become restricted. Such entities are competitive with respect to arrival and exploitation, i.e. in terms of "primary resource capture" (see Chapter 1). They also have relatively unspecialized requirements for resources and little need to adjust to heterogeneous conditions other than by reproducing. Regenerative processes are therefore emphasized, allowing very rapid rates of proliferation and dissemination of reproductive units through space and time. Variation produced by recombinatorial, developmental or behavioural mechanisms (Chapters 4, 5) tends to be minimized. Consciously or unconsciously, this is the pattern that has underlain much of the short-term thinking that is at the base of current notions of evolutionary, ecological and economic success.

Entities with attributes favoured by R-selection are known by many names—some disparaging, others admiring. Commonly they are described as "ephemerals" or "opportunists". In plant communities they are often known as "weeds" or "ruderals" (hence the term "R"-selection). In a social context, they may be referred to as "fly-by-nights", "cowboys", "raiders" or, more approvingly, as "entrepreneurs" or "pioneers".

2.2. Combatants and Collaborators

As conditions stabilize following a disturbance, an increasing number of entities has the potential to gather within the same arena and deplete resources, to the point where their contextual boundaries coincide (Chapter 1, Fig. 1.1). After the goldrush, increased access to resources by any one entity or group can only be achieved by means of trade or takeover. The ability of each entity to retain or gain resources now depends on mechanisms affecting the self-integration and degeneration of contextual boundaries. These mechanisms and their relation to genetic similarities and differences will be discussed in more detail in Chapter 7. For the moment, their main significance lies in the fact that they result in the "combative" and "collaborative" ecological strategies of entities that develop in circumstances where there is a potentially high incidence of competitors. Such circumstances cause "C"-selection.

Since C-selected entities inhabit heterogeneous environments over relatively long time intervals, during which they may have a wide variety of close encounters with others, they tend to possess a high degree of versatility. This versatility is associated with an emphasis on conversional processes—which permit retention of resources, and distributive processes—which enable invasion of hostile territory. Recycling processes that enable redistribution to defensive or invasive "battlefronts" are often also important. The formation of co-operative networks and partnerships allows new capabilities that can also enhance combative prowess. C-selected entities can be expected to be genetically variable at the population level of organization and developmentally or behaviourally variable at the individual level.

2.3. Stress-tolerators

Stress or adversity can be thought of as any more or less continuously imposed kind of environmental extreme, other than a high incidence of competitors, which inhibits the proliferation of the majority of entities under consideration. Particularly in terrestrial environments, many stress factors may ultimately operate by compromising the mechanisms that circumvent or protect from oxygen toxicity (see chapter 5).

Those relatively few "S"- or "A"-selected entities which can tolerate, or indeed develop best in the relative absence of competitors under stressed conditions, do so because they have specialized attributes. For example, organisms inhabiting deserts have attributes which protect them from desiccation and extremes of temperature.

S-selection, like C-selection, is a feature of relatively stable environments (hence both C- and S-selection represent different aspects of K-selection—see Chapter 4) and especially favours attributes associated with protective, conversional processes. Co-operative interactions and distributive processes that allow emigration can also be an asset. Purely S-selected entities, in being highly specialized, tend to lack versatility. However, S-selection is often associated with individual life cycle stages or alternative phenotypes of versatile organisms.

3. Corporate Strategies and Collective Organization

It takes all kinds to make a world, the saying goes. Correspondingly, interactions between pioneers, combatants, collaborators and specialists occur in all kinds of dynamic assemblages. These interactions can result in remarkably flexible and coherent responses to environmental circumstances. As the balance between exploration, exploitation, conservation and redistribution in the assemblages shifts, a series of changes in composition occurs until either a relatively stable, "climax" phase is attained or degeneration ensues. From the outside, these changes may seem to be due to "executive decisions", but they actually follow automatically from the interplay of feedback processes at dynamic boundaries.

3.1. Development of Indeterminate Individuals

Even in determinate organisms, the processes that lead to the development of an "adult" state are due to initial proliferation of like entities followed by subdivision into specialisms interconnected by infrastructural networks (Chapter 4). There is therefore a progressive shift in balance from differentiation to self-integration and final degeneration. Transitions between these stages may either be continuous or—as in some metamorphosing forms—abrupt (Chapter 5). Although the environment may affect the timing and scale of these transitions, it cannot alter their sequence or outward expression and so changes in body pattern over a lifetime are fundamentally pre-programmed. Developmental stages are attuned with life history rather than local circumstances.

In indeterminate organisms, changes in developmental or behavioural mode are more closely attuned to local circumstances and so each mode may be considered to correspond with a particular ecological strategy. Indeterminate organisms therefore have the potential to combine a variety of attribute sets, corresponding with an equivalent variety of ecological strategies, over their indefinite life spans. Nonetheless, it is commonly found that the attribute sets of particular organisms are biased towards particular strategies. In extreme cases, entire sets of attributes may be excluded from the life span; in particular the dispersal phases of R-selected modes and the protective and invasive modes of S- and C-selected modes may be omitted.

Typically, the first phase of development of indeterminate organisms, whether it follows the germination of a seed, a spore or other kind of propagule, involves the generation of some energy-gathering structure. During this "germling" or "seedling" phase the organism is relatively uninsulated and so unprotected against hostile environmental factors and deprivation of resources. Such phases are therefore unestablished and vulnerable. They possess R-selected attribute sets, and in some plants and fungi, notably annual weeds and asexually reproducing "moulds" may represent the only or predominant phase in the life span prior to the production of seeds and spores.

In other cases these relatively uninsulated phases are superseded by self-integrated phases equipped to explore, invade and retain resources. The emphasis therefore shifts towards C- and/or S-selected attribute sets. The latter are associated with the formation of protective barriers such as tree bark and tough "rinds" of mycelium (pseudosclerotial plates), as well as distributive structures such as stolons, acutely branched hyphae, mycelial cords and rhizomorphs.

In many fungi, the changeover from relatively uninsulated to self-integrated states is often associated with a shift in the pattern of reproduction. Relatively uninsulated states often produce an abundance of asexual spores. The "sporophores" on or from which these spores are formed are usually small—often microscopically small—and so do not need large supplies of resources. More self-integrated states, on the other hand, produce larger, sometimes huge, sporophores—brackets, morels, puffballs, toadstools and the like—in which the spores are typically produced sexually (i.e. by meiosis). These sporophores obviously represent a much greater investment: although their energy cost may be offset to some extent by their more effective insulation than asexual sporophores, they draw off considerable supplies. They are therefore only produced once a mycelium has gained access to a sufficient pool of resources and can redistribute them through a networked structure. Generally speaking, the size of sexual sporophores produced by fungi is quite a good indication of the resource pool its mycelium can draw on. Compare, for example, the tiny toadstools of *Marasmius epiphylloides* growing on an ivy leaf with the huge clump of toadstools produced by *Armillaria bulbosa* produced on a tree stump (Fig. 6.6).

In order to provide an idea of how the varied organizational modes of indeterminate systems are attuned to circumstances that are unpredictable at specific locations, a number of fungal case studies will now be described.

The first of these examples is a bracket-forming fungus, *Heterobasidion annosum*, which causes considerable damage in northern coniferous forests, killing and rotting the roots of pine trees and decaying the heartwood of larches, spruces and firs.

A large amount of the damage caused by *H. annosum* is due to the ways in which coniferous forest has been exploited as a source of commercial timber. As shown in Fig. 6.7, the fungus gains access to the wood of trees through wounds and cut stump surfaces produced as a consequence of felling and extracting timber.

Figure 6.6. Fruit bodies of *Marasmius epiphylloides* on an ivy leaf and *Armillaria bulbosa* on a tree stump. (Photographs by Ed Setliff and John Webster)

Air-borne basidiospores of the fungus arrive at newly exposed wood surfaces and germinate to initiate colonization. From the perspective of the fungus, exposure of the wood surfaces constitutes an "enrichment disturbance", and the first phase of colonization is predominantly proliferative. During this phase the mycelium is relatively uninsulated and characteristically may produce large numbers of asexual spores ("conidia") from microscopically small sporophores ("conidiophores"). Eventually, once mating has occurred, a secondary, relatively more insulated, mycelial phase (see Chapter 7) becomes established. The boundaries of this phase are often protected by the development of dark, watertight rinds ("pseudosclerotial plates"), and on the outside of the tree or stump, large, well-insulated, bracket-shaped sexual sporophores are formed. Within these sporophores, meiosis leads to the generation of innumerable basidiospores that drop down through the tubes of the spore-producing surface ("hymenium") into turbulent air and are carried away to new locations.

Although dispersal of air-borne spores gives rise to new infection centres, a more immediate concern is the fact that the mycelium can spread across contacts between the roots of infected trees or stumps and the roots of neighbouring, healthy trees (Fig. 6.8). This mode of transmission is in fact common amongst root-rotting fungi, and an analogous process can occur amongst certain stem-infecting fungi, such as *Hymenochaete corrugata* (Fig. 6.9), providing that the "bridging mycelium" is adequately protected from desiccation (and/or associated oxidative stress).

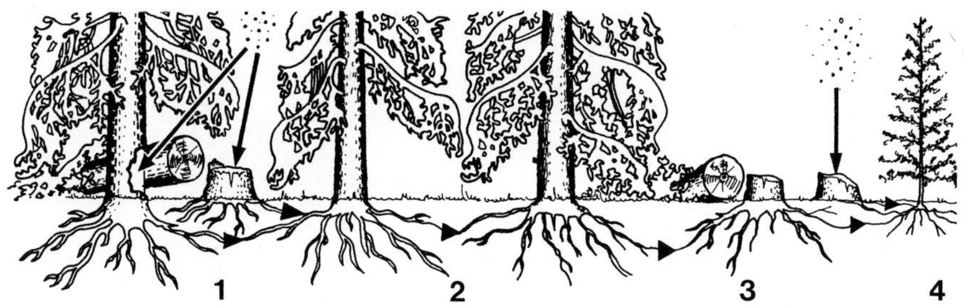

Figure 6.7. How the root disease-causing fungus *Heterobasidion annosum* becomes established and spreads in conifer plantations. Stages shown from left to right are as follows. (1) *Thinning*. Spores land and germinate on wounds and on recently exposed stump surfaces. (2) *Continuous*. The fungus transfers across root contacts between standing trees. (3) *Final felling*. Decay occurs in stumps of trees that have already been infected as well as stumps that become freshly colonized by spores. (4) *Regeneration*. Infections established in the previous stand are transmitted to young trees via root contacts. (From Stenlid, 1986).

Once the secondary phase mycelium of *H. annosum* has transferred across a root contact, further progress depends on whether it gains access to the dead inner core of heartwood or produces a superficial covering of mycelium over the root surface. The importance of this superficial mycelium in infection of pine roots is indicated by the fact that trees grown on acid soils, where growth of this mycelium is inhibited, are relatively non-susceptible to attack. Pine trees grown on alkaline soils, where the superficial mycelium can be present well in advance of infection of the wood cylinder, are extremely susceptible to attack.

The role of the superficial mycelium in infection can be understood in terms of the interaction between fungus and tree as hydrodynamic systems. Conditions within the functional (water-conducting) sapwood of living trees are inimical to extensive fungal proliferation, probably largely because the high water content limits diffusion of oxygen.

The ability of many fungi to develop vigorously in the wood of trees therefore depends on the dysfunction and resultant aeration of sapwood. Dysfunction can be brought about either by degenerative processes, such as those which result from heartwood formation or internal competition for resources (see Chapter 5), or by external agencies such as damage or drought. However, the spread of such dysfunction is commonly limited by hydrophobic barriers produced by living tree cells. These barriers, which may themeselves be interpreted as a response to oxidative stress (see Chapter 5) therefore limit the availability of dysfunctional wood for colonization by fungi, and in addition may contain chemicals which are inhibitory or toxic to fungi.

Figure 6.8. Superficial white mycelium (arrowed) of *Heterobasidion annosum* spreading from an infected onto a healthy root. (Courtesy of J. Stenlid).

Figure 6.9. Mycelial bridges (arrowed) formed by the fungus *Hymenochaete corrugata* across contacts between infected and uninfected hazel (*Corylus avellana*) stems (left) and between a hazel stem and *Clematis* vine (right). (From Ainsworth and Rayner, 1990).

Actively pathogenic fungi, such as *Heterobasidion annosum*, bring about the dysfunction of sapwood themselves (see Chapter 7). One way in which they can achieve this is by producing non-assimilative, well-insulated mycelial phases which are relatively impermeable to any toxic chemicals, whilst being able to produce extracellular compounds that kill host cells and destroy barriers. This is how the superficial mycelial phase, and its ability to pave the way for assimilative hyphae, can be viewed. As will be explained further below and in Chapter 7, the production of non-assimilative mycelial phases is also a key feature of fungi with combative ecological strategies.

The case of *H. annosum* therefore demonstrates the usefulness of a versatile repertoire that allows the fungus to perform the predictable tasks required for successful colonization at the unpredictable locations that interaction with a dynamic habitat (the tree) requires.

In *H. annosum*, colonization is initiated following a disturbance, and so the first developmental phase exhibits attribute sets appropriate to R-selection; other wood-inhabiting fungi establish under circumstances imposing S- or C-selection. With regard to S-selected modes, two interesting cases are provided by what have been described as "heartrot" fungi and "specialized opportunists".

Heartrot fungi colonize the dysfunctional tissues of heartwood, but in so doing have to contend with inhibitory phenolic and terpenoid substances as well as a gaseous regime liable to be rich in carbon dioxide and containing little oxygen. They are clearly stress-adapted in that they grow slowly, lack combative ability and reproduce only sporadically or after a long time, whilst generating individual systems that can eventually become very extensive and support large sporophores.

Specialized opportunist fungi are able to establish themselves within functional sapwood. They rely on the fact that although the high water content of sapwood is inimical to the development of actively assimilative mycelial phases, limited development of benign, suitably specialized modes is possible. When dysfunction occurs, the fungi switch into more actively assimilative states and are well placed to take in available resources before competitors establish. They therefore switch from stress-adapted to ruderal modes, and subsequently may progress further into combative states.

An increasing number of fungi are thought to colonize trees in this way, including such familiar examples as the birch polypore, *Piptoporus betulinus*, and King Alfred's cakes, *Daldinia concentrica* (Fig. 6.10). Similar kinds of behaviour are shown by some animal-infecting fungi, which are benign most of the time but switch into more active states if their host is stressed in some way, for example by trauma, disease or suppression of the immune system. A well known case in human beings is provided by *Candida albicans*, which causes oral and vaginal "thrush" as well as more severe, sometimes fatal, internal disease.

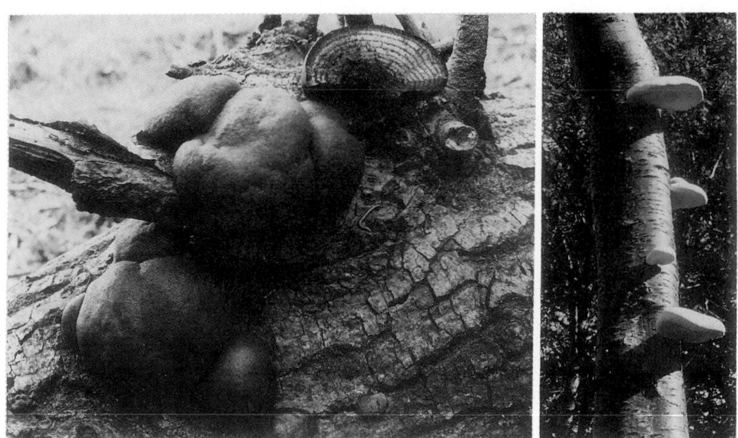

Figure 6.10. Specialized opportunists: King Alfred's cakes, *Daldinia concentrica*, on an ash log (one fruit body has been cut open to show the concentric zones of water-storage tissue within) and (right), the birch polypore, *Piptoporus betulinus* on a birch trunk. (Photographs by John Webster)

A variety of fungi inhabiting the woodland floor can establish themselves in pieces of wood which have fallen from the canopy and are hence already likely to be undergoing decay and inhabited by other fungi. These fungi commonly produce mycelial cords, and include the examples illustrated growing in matrices in Figures 6.2 to 6.4. The cords provide the fungi with a means of exploring between discontinuously distributed sources of nutrients and produce patterns that correspond with a variety of "foraging strategies". The latter depend partly on the genetic make-up of the fungi themselves, and partly on environmental circumstances.

Short-range foraging, as illustrated by the sulphur tuft fungus, *Hypholoma fasciculare*, involves highly concerted processes and is most effective when colonizable resources are likely to be encountered close to a current site of assimilation (Fig. 6.11). Dense outgrowth, associated with a large number of anastomoses, is superseded on contact with a suitable nutrient source by considerable redistribution from non-connective to connective mycelium, especially once exploration is resumed. Long-range foraging, as shown by *Phanerochaete velutina*, is more individualistic, with sparser, less interconnected cords and less evidence of redistribution following arrival at a nutrient source (Fig. 6.12). It is characteristic of fungi which habitually colonize widely separated resources; however, these fungi may still be capable, as is *P. velutina*, of producing short range patterns if they arrive at nutrient sources that are disproportionately large (Fig. 6.13).

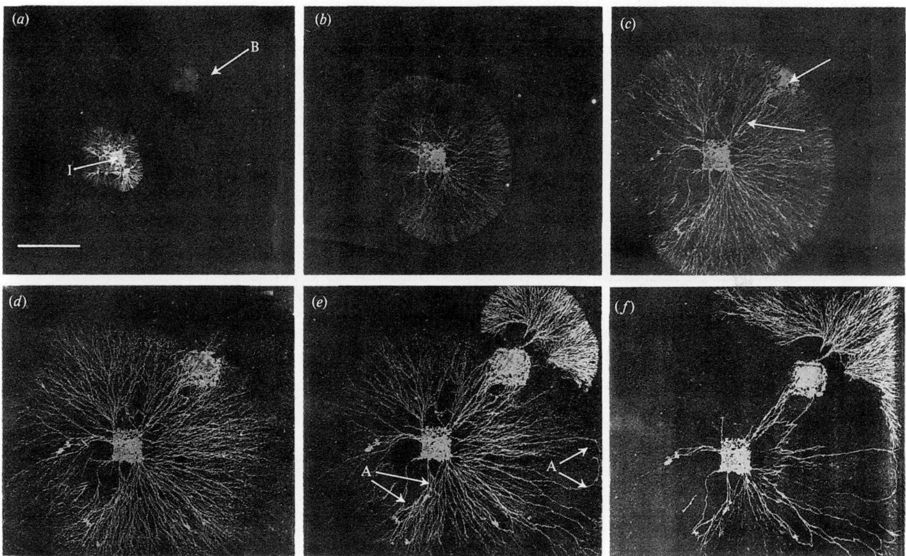

Figure 6.11. *(a-f)* Stages in short range foraging by the sulphur tuft fungus, *Hypholoma fasciculare* between blocks of wood in soil. The experimental design is as in Fig. 3.3. I, inoculum; B, bait; A, anastomoses resulting in persistent networking. Bar represents 4 cm. (From Dowson et al., 1986)

The extent to which cord-forming fungi are genetically predisposed to exhibit short- or long-range foraging may be related to the degree to which they are subject to R-, C- and S-selection. Although all are combative relative to other wood-inhabiting fungi, relative to one another they can still be allocated different positions in the R-C-S continuum. Relatively R-selected forms are most liable to forage over short range, and to switch most readily from explorative into reproductive or assimilative states. Of the examples shown growing in matrices (Figs 6.2 to Fig. 6.4), *Coprinus radians* is relatively R-selected, *Phallus impudicus* and *Coprinus picaceus* relatively C-selected, and *Trechispora vaga* both S- and C-selected.

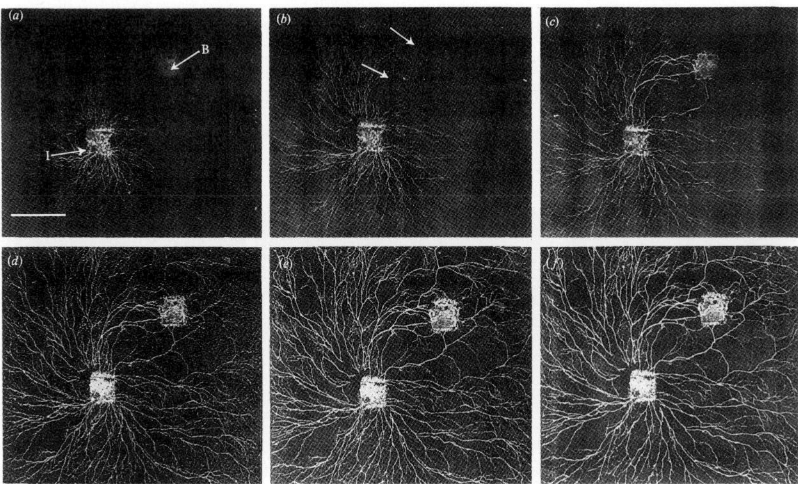

Figure 6.12. (*a-f*) Stages in long range foraging by the fungus, *Phanerochaete velutina*, between wood blocks in soil. Arrows indicate curvature of mycelial cords towards the bait. Bar represents 4 cm. (From Dowson et al., 1986)

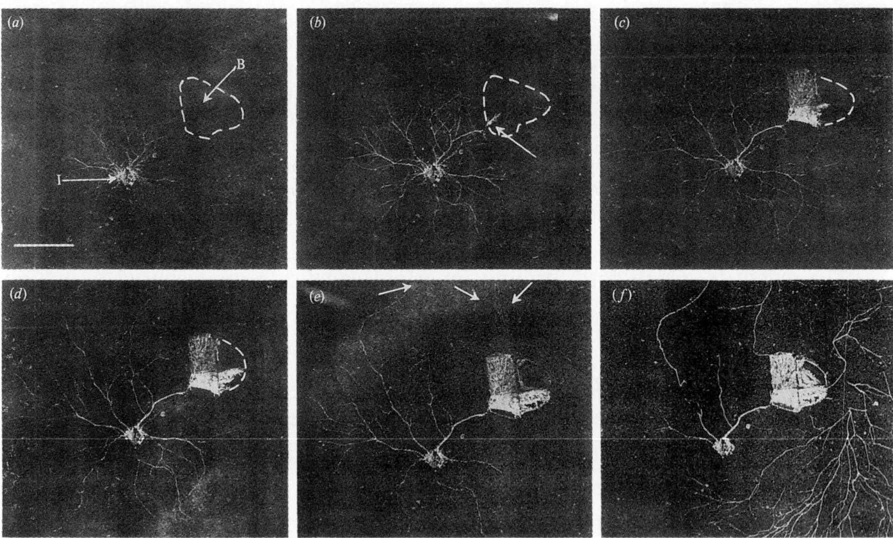

Figure 6.13. (a-f) Stages in short range foraging pattern shown by the fungus, *Phanerochaete velutina*, when grown between small and large woodblocks in soil. Arrows show diffuse mycelium at point of contact with bait (b) and redirected growth of a cord towards the bait (e). Bar represents 4 cm. (From Dowson et al., 1986)

The occurrence of different foraging strategies has also been used to account for the varied branching patterns of plant roots, stolons and rhizomes. In particular, two extreme patterns have been recognized, phalanx and guerilla, the former resulting in dense proliferation over a short range, and the latter with more invasive, long-range growth. However, the concept of foraging strategies arose neither from observations of fungi nor of plants, but rather from organisms that at the individual level are all too determinate.

3.2. Social Development

Figure 6.14 illustrates the patterns formed by the raid fronts of three different species of army ants which characteristically feed on rather different kinds of prey. The similarities of these foraging patterns to those of fungi and plants—and the way they are influenced by the frequency and size of "resource depots"—are striking, demonstrating how organizational processes can transcend classificatory distinctions. Furthermore, when the entities within these social systems are examined more closely, it becomes clear that notwithstanding their genetic similarities, they are by no means all alike to one another. Rather, as explained in Chapter 2, they consist of "castes" which play distinctive roles in the gathering, redistribution, conservation and defence of resources, and the allocation of these resources to growth of the colony or reproduction.

Determination of the castes in social insect colonies is generally based on the way the larvae are nurtured—i.e. fed on particular diets by "nurse" workers. However, in one genus of stingless bees there is evidence for superimposition of genetic control.

In some cases the workers may be differentiated into subcastes that either have distinctive anatomies or exhibit shifts in physiology and/or behaviour as individuals age. The generation of anatomically distinctive castes or "morphs" involves what has been termed "allometry"—varied expansion of the boundaries of different body components (Fig. 6.15). Differentiation over time is exemplified by the switchover from inside the nest to outside the nest activities in workers of the European wood ant, once they are more than 50 days old.

The castes therefore represent alternative phenotypes with roles equivalent to differing ecological strategies. Also, as with the development of indeterminate individuals (see above) there is a general shift in balance from proliferative processes to more specialized, conservative and combative functions as the societies mature.

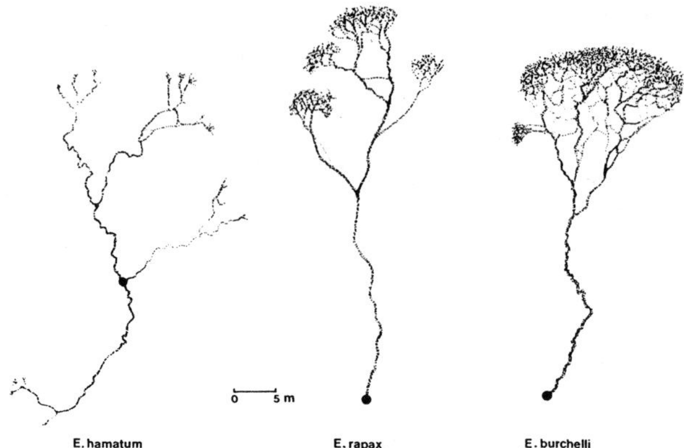

Figure 6.14. Foraging patterns formed by the raid fronts of three different species of army ants in the genus *Eciton*. *E. hamatum* preys on widely dispersed social insect colonies and exhibits a sparsely branched, long-range pattern. *E. burchelli* preys on social insects and evenly distributed large arthropods and produces a delta-like swarm front with many branches and anastomoses. *E. rapax* has an intermediate diet, mostly eating social insects but concentrating rapid attack on their nests to capture both adults and brood; it also captures some large arthropods. It has been shown that such differences in pattern can be generated by small differences in the relative attractiveness of the trail pheromones that enhance the coherence of the movements of individual ants. (From Burton and Franks, 1985, in part after C.W. Rettenmeyer).

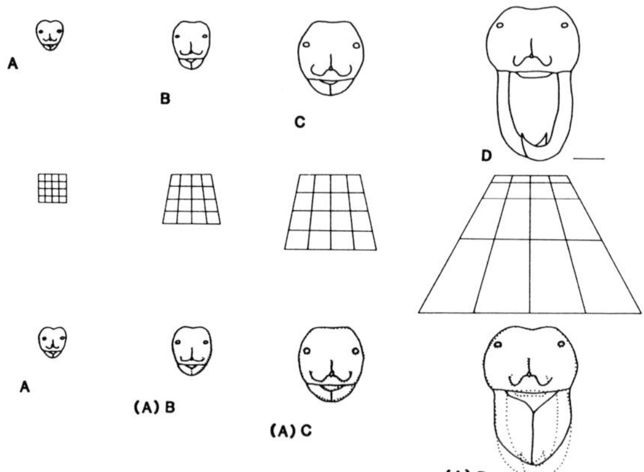

Figure 6.15. The generation of different castes of army ants by differential expansion of their body boundaries (allometry). The top row A, B, C, D are the heads of an *Eciton burchelli* minim, medium worker, submajor and major respectively. The grids below each caste represent the form of cartesian transformation required to modify the minor caste into each of the larger forms. A(B), A(C), A(D) are the modified minor outline (solid line) superimposed on the dotted outline of the respective larger caste. The scale bar represents 1 mm. (From Franks and Norris, 1987).

The above outline of roles and shifting balances in insect societies is strongly reminiscent of those shown in other societies, not least those of human beings. That social insect systems should long have been considered to provide perhaps the most sophisticated biological models of human organizations may therefore come as no surprise.

However, there are two dangers in relating social insect systems to human societies.

Firstly, there is the danger of imposing human preconceptions of social order onto the observed systems. Correspondingly, the descriptions of worker, queen and soldier castes have more than a few overtones of human notions of labouring, royal and military components of society. This is a general danger, and one which I have been only too conscious of throughout this book. As in all cases when relating biological knowledge to human concerns, it is important, but difficult, to try to observe living systems without prejudice. Only once (and if) recurrent patterns have become evident is it then appropriate to consider whether they can helpfully be related to our own organizations.

Secondly, there is the danger of becoming so focused on one particular, rather obvious, model system that other less obvious, less prejudicial and hence potentially more generically instructive parallels may be overlooked. This is why I have tried in this book to look at the broad sweep of biological patterns across the mechanistic and organizational divides.

3.3. Community Development

When patterns within natural biological communities—assemblages of different kinds of organisms within ecosystem boundaries—are examined, the themes and shifting balances apparent in indeterminate and social development again come into view. Indeed, ecological strategies theory was developed primarily as a result of trying to understand processes leading to stability and instability in natural communities rather than at smaller organizational scales.

This commonality of themes across organizational divides emphasizes the usefulness of thinking about all kinds of dynamic assemblages—cells, colonies, multicellular organisms, societies or communities—in terms of the way they gather, conserve, distribute and redistribute energy within dynamic boundaries.

The interplay between differentiation and integration during the development of community organization was highlighted by a debate in the early part of the twentieth century between "individualistic" and "organismal" schools of thought.

The organismal school was represented by Clements, who envisaged that processes of "succession" in land plant communities are directed towards a final, stable, "climax" or equilibrium state determined primarily by prevailing climatic conditions. The community in this climax was regarded as a self-sustaining functional whole characterized by a multiplicity of interdependencies between its "co-evolved" components. As such, the climax community was considered to be comparable to a kind of "superorganism" whose parts are determined by the whole—by top-down, not bottom-up management. A more recent manifestation of the same kind of thinking, known as the Gaia hypothesis views not just particular biological communities but the entire planet earth as a self-perpetuating, self-correcting superorganism.

It follows from the organismal school of thought that different kinds of plant communities can be recognized and classified in terms of characteristic "associations". These associations consist of individual plant species that show high "constancy" (ever-presence) and/or "fidelity" (exclusivity) in a particular community type. Correspondingly, "floristic analysis" became (and still is) a widespread research tool in the description of plant communities. Also, since plants are, by dint of their photosynthetic ability, "primary producers", i.e. the energy gatherers of self-sustaining ecosystems, they provide both the hub and physical architecture around, upon and within which the activities of other organisms revolve and ramify. These other organisms may also form characteristic associations.

The individualistic countercurrent to organismal thinking was provided by Gleason who considered that the species composition of plant communities is determined by the random mixing of organisms with attributes or traits that allow them to thrive in the same neighbourhood. From this viewpoint, any apparent order in the system arises from chance associations of individuals rather than some primary drive for co-operation. In other words: neighbours are the product of random assortment; the whole is equal to the sum of the parts; groups are determined by individuals, not individuals by groups; differences in initial composition of communities lead to proportional differences in final composition; there is bottom-up, not top-down management.

The view maintained throughout this book is that such extreme versions of the organization of dynamic systems are interdependent polarities, not independent alternatives. There is top-down *and* bottom-up management. Individuals specify group structures *and* groups impose selection on individuals. Whereas the selection *of* groups through the sacrifice of individuals is illogical, selection *by* groups is a common and important phenomenon, often referred to in human communities as "peer pressure". Systems determine boundaries *and* boundaries define systems. The balance between differentiation *and* integration shifts as boundaries become more open *and* as they become more closed. Common, unidirectional patterns of community change can be identified, *and* small differences in initial composition can lead to *both* small *and* large differences in final composition.

One way in which individualistic and collectivistic viewpoints can be reconciled into a single, fluid perspective of processes of community equilibrium and change, with infinite scope for variations upon a theme, is illustrated by the schema in Fig. 6.16. The schema draws on ecological strategy concepts as a means of characterizing the changing attributes of community members over time. It also identifies four fundamental processes, allied to those described earlier in this chapter, that govern community stability and instability. These processes are community opening, community closure, stress alleviation and stress aggravation.

The development of a biological community can in many ways be likened to the generation of a city, not—as has been the case with certain recent, soulless monstrosities—according to some preconceived plan, but rather from small beginnings and through locally unpredictable feedbacks. Upon or within some base or container, both superstructure and infrastructure are assembled by indeterminately expanding organizations and their offshoots, be these of plants, fungi or colonial animals. Meanwhile, the lives of mobile, determinate beings weave in, out and through these structures.

Some form of boundary opening or discontinuity is vital for recruitment into any kind of community. Generally, this takes the form of a disturbance and so, in the absence of stress, favours rapidly proliferating organisms or states. Correspondingly, the first builders of superstructure in terrestrial plant communities establishing on virgin soil, are often annual weeds in such families as the cabbage family (Cruciferae) (e.g. shepherd's purse, *Capsella bursa-pastoris* and hairy bittercress *Cardamine hirsuta*) and campion family (Caryophyllaceae) (e.g. chickweed, *Stellaria media*). Similarly, in fungal communities establishing amongst newly available supplies of organic nutrients, rapidly proliferating "moulds" such as *Mucor* and *Penicillium* are often first to dominate the scene.

If, on the other hand, initial conditions following disturbance are stressful, then the pioneers may combine both R- and S-selected attributes. For example, on the mud at the leading edge of saltmarshes, glasswort (*Salicornia* spp.), a succulent (cactus-like) member of the spinach family (Chenopodiaceae) predominates in the physiologically dry (because of salt water) conditions. At the exposed margin of sand dunes resilient Chenopods such as Hastate Orache (*Atriplex hastata*) and prickly saltwort (*Salsola kali*) occur alongside Crucifers such as sea rocket (*Cakile maritima*). Similarly, in the actually or physiologically dry conditions within cereal grain stores and pots of jam, prolific spore-producing species of *Aspergillus*, which are able to grow at low water potential, are often the first fungi to become established. On the other hand if conditions are highly inimical to growth, organisms capable of surviving or developing as dormant or quiescent states may eke out an existence, ready to spring into action when amelioration occurs.

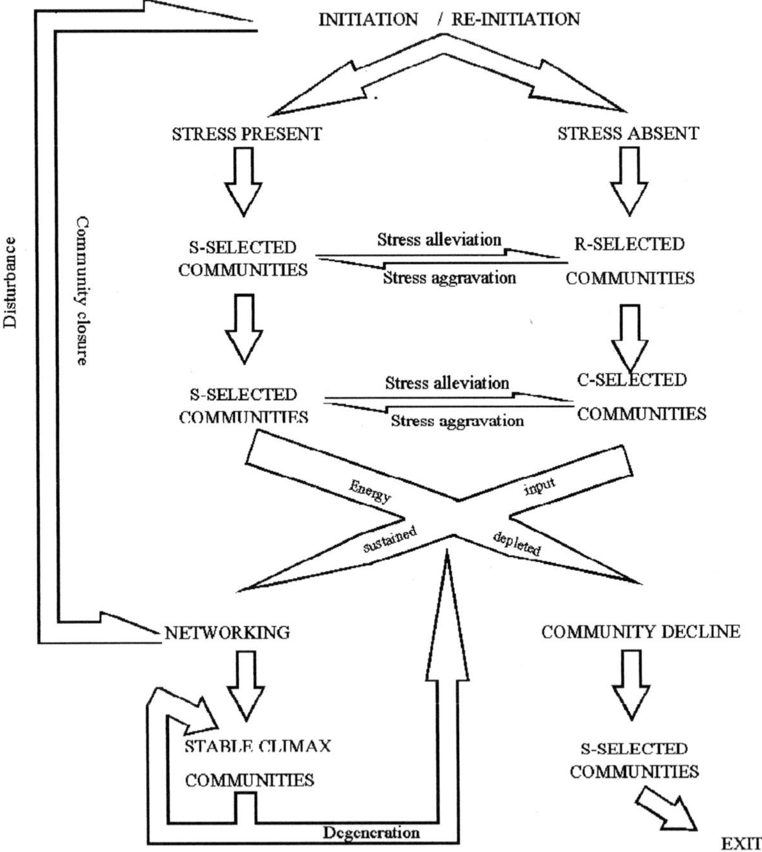

Figure 6.16. Schema showing processes that lead to stability and change in natural communities of organisms. (Drawn with help from R.E. Guevara)

Once community development has been initiated, further recruitment and proliferation of pioneers leads inevitably to community closure due to limitations in space or resource supplies. As this happens, distributive infrastructures begin to develop. For example initially separate plants may come to be interconnected by the hyphae of mycorrhizal fungi. Interestingly the latter fungi do not invade the roots of pioneer members of the Chenopodiaceae, Cruciferae and Caryophyllaceae but do invade those of most other plant families.

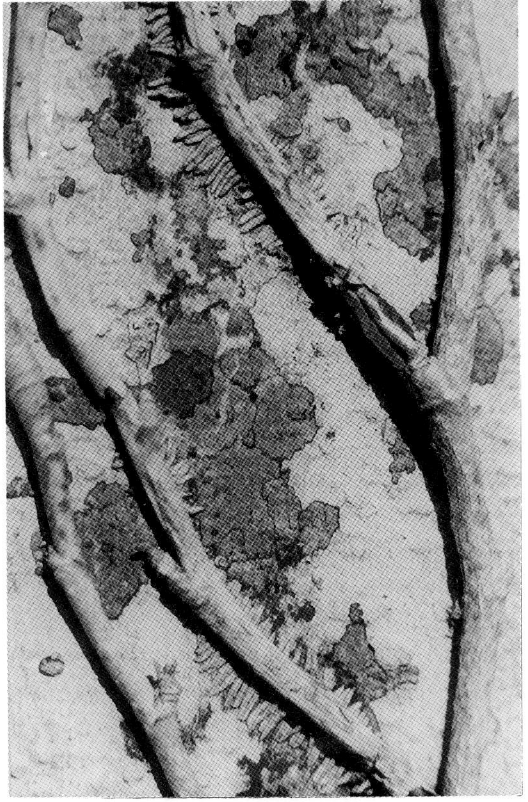

Figure 6.17. Part of the surface of a sycamore trunk that has been colonized by a white lichen, possibly *Lecanora chlarotera* and a grey-green lichen, *Enterographa crassa*, which consists of many individuals delimited by narrow dark lines. Stems of ivy (*Hedera helix*) are using *E. crassa* as "stepping stones" across the white lichen which inhibits the ivy from forming the adhesive roots that enables it to climb. (Courtesy of Linda Tong and Julia Brixey).

In the relative absence of stress, the potential for combat intensifies with community closure. As this occurs, pioneers must either vacate the scene or possess or develop attributes which resist the territorial advance of their neighbours, and any new recruits must be able to oust residents. Possession of resilient boundaries and associated abilities to poison, penetrate into and overtop neighbours becomes the order of the day, culminating either in complex stalemates or the rise to dominance of one or a few entities. Lichen communities developing over the surfaces of sea shells, boulders or tree trunks exemplify the resulting patterns on spatially small, time-extended scales (Fig. 6.17). The first colonizers are typically "crustose" forms that form a thin, light-gathering layer over (and sometimes slightly below) the exposed surfaces. Over time, the more robust of these start to come into contact, but fail to coalesce, leaving map-like boundaries at their interfaces. Such mosaics may persist for many decades. However, they get overtopped if leafy or "foliose" or shrubby "fruticose" forms can gain a foothold. Similar processes occur within purely fungal communities (Fig. 6.18). In plant communities, scattered annuals are superseded in turn by ground-covering and then increasingly shrubby and tree-like forms. The resulting increased scale and complexity of the resulting fractal architecture then provides within itself the theatre for self-similar series of community plays at diminishing scales.

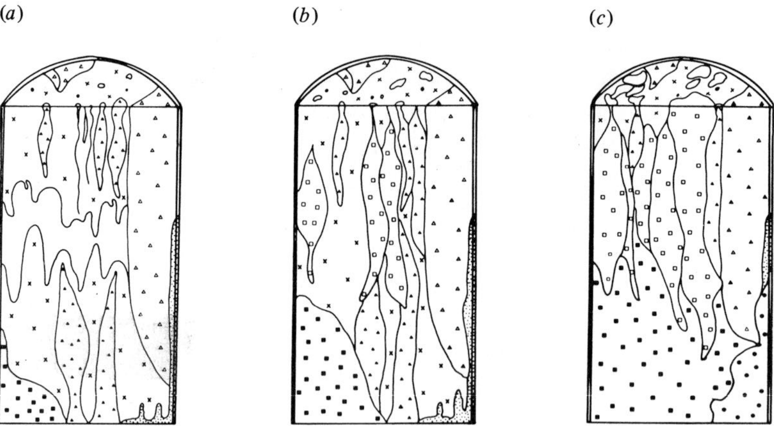

Figure 6.18. Diagram illustrating a typical pattern of development of a community of fungal mycelia in a beech log of the kind shown in Figure 2.21. Patterns of community organization 6 months (*a*), 12 months (*b*) and 18 months (*c*) after the log was cut are shown. Symbols used to represent the different fungi present are as follows: x, stained or discoloured wood containing microfungi and/or the basidiomycetes *Chondrostereum purpureum* and *Corticium evolvens*; solid triangles, the ascomycete, *Xylaria hypoxylon*, colonizing from air-borne or soil-borne spores; open triangles, members of the ascomycete genus, *Hypoxylon*, colonizing as

"specialized opportunists" already present in the wood before it was cut; open squares, combative air-borne basidiomycetes such as *Coriolus versicolor*, *Bjerkandera adusta* and *Stereum hirsutum*; solid squares, combative, early-arriving mycelial cord-forming basidiomycetes, e.g. *Phallus impudicus* and *Tricholomopsis platyphylla*; solid circles, combative, late-arriving cord-formers, e.g. *Phanerochaete velutina*; stipple, the rhizomorph-forming basidiomycete, *Armillaria bulbosa*. (Modified from Coates & Rayner, 1985).

Stress aggravation and stress alleviation respectively worsen or ameliorate the imposition of extreme conditions, and can be brought about by changes in circumstance initiated both within and outside communities (i.e. by both endogenous and exogenous influences). Depending on its timing in relation to overall community development, stress alleviation may shift the balance towards either R- or C-selection, whereas stress aggravation at any stage enhances S-selection.

The importance of stress alleviation in community development has already been illustrated by the specialized opportunist fungi colonizing debilitated host organisms, and by the blooming of deserts after rain. More generally, there are numerous examples where pioneer organisms pave the way for others to follow by obviating selective barriers and extremes. Crustose lichens produce acids which etch into rock surfaces, hastening processes of natural weathering which generate irregularities and soil in which plants can become established. The annual plants at the front of sand dune systems act as foci around which blown sand and other debris accumulates. This allows the marram grass, *Ammophila arenaria* and its sand-trapping stolon and root systems and attendant mycorrhizal fungi to establish, accompanied by the sand sedge, *Carex arenaria*. These plants in turn provide shelter in which ground-covering mosses, grasses and plants such as the restharrow (*Ononis repens*) establish and produce a stable surface that allows shrubby plants and finally trees to establish (Fig. 6.21). Similarly, in fungal communities, active pathogens causing dysfunction of their hosts allow more combative organisms or modes to follow, and the metabolic activity of *Aspergillus* species in grain stores potentiates colonization by less desiccation-tolerant forms.

External factors causing stress aggravation include climate change, pollution, herbivore grazing and soil erosion. Internal factors include the depletion of essential or easily assimilable nutrients, the accumulation of toxic waste products and the generation of high temperatures through metabolic activity in a confined space such as a grain store or compost heap. With regard to the depletion of nutrients, an explanation that was once widely favoured for community change in fungal communities was that progressively more refractory sources of carbon, such as sugars, cellulose and lignin, would be used in sequence by successively more specialized organisms. However, it has subsequently been recognized that whilst such changes in nutrient availability play an important part in fungal community dynamics, it should not be the only or even predominant consideration.

To summarize, by a variety of routes, and spread over what may be several or more cycles, the scene is set for the establishment of multilayered, multifaceted, closed community structures. These structures are dominated by organisms that are combative or specialized relative to previous residents and increasingly prone to develop interconnections and interdependencies. Providing that energy input to the community is sustained, then a kind of dynamic equilibrium state, or climax, may be attained. However, this state can perhaps never be so fully integrated as to cause neighbours to occur in absolutely predictable arrays, or to resist forever those processes that lead to degeneration or radical shifts in organizational pattern. Also, whenever energy inputs are not sustained, as in some kinds of fungal community, the continued depletion of resources leads inevitably to degeneration and decline.

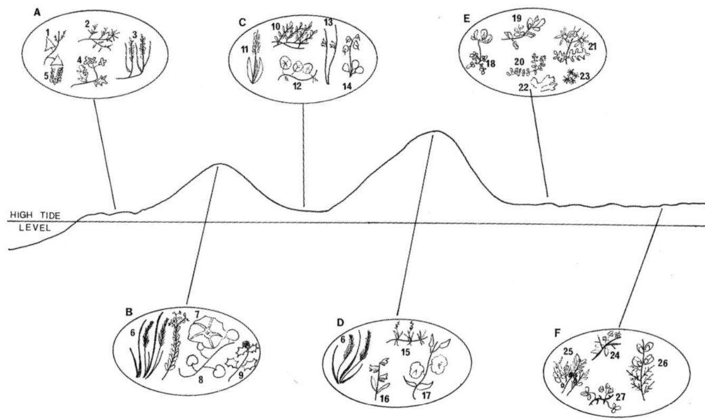

Figure 6.19. Diagram illustrating plant community succession on British sand dunes. A, foredunes; B, very mobile dune; C, dune slack; D, mobile dune; E,F, fixed dunes with full plant cover and dune scrub. Plants typical of different stages are indicated by numbers as follows: 1, hastate orache, *Atriplex hastata*; 2, prickly saltwort, *Salsola kali*; 3, sand couch, *Agropyron junceiforme*; 4, sea rocket, *Cakile maritima*; 5, sea purslane, *Honkenya peploides*; 6, marram grass, *Ammophila arenaria*; 7, sea spurge, *Euphorbia paralias*; 8, sea bindweed, *Calystegia soldanella*; 9, sea holly, *Eryngium maritimum*; 10, creeping willow, *Salix repens*; 11, marsh orchid, *Dactylorchis praetermissa*; 12, marsh pennywort, *Hydrocotyle vulgaris*; 13, rushes, *Juncus* spp.; 14, large wintergreen, *Pyrola rotundifolia*; 15, sand sedge, *Carex arenaria*; 16, viper's bugloss, *Echium vulgare*; 17, evening primrose, *Oenothera* spp.; 18, bird's foot trefoil, *Lotus corniculatus*; 19, restharrow, *Ononis repens*; 20, thyme, *Thymus* spp.; 21, ragwort, *Senecio jacobea*; 22, dog lichen, *Peltigera canina*; 23, moss, *Tortula ruralis*; 24, blackthorn, *Prunus spinosa*; 25, sea buckthorn, *Hippophae rhamnoides*; 26, gorse, *Ulex europaeus*; 27, bramble, *Rubus fruticosus*.

None of the pattern-generating community interplay that has just been outlined would be possible if it were not for the capacity of the multifarious forms of living beings to open up and seal off their contextual boundaries to and from one another. This capacity, above all others, generates and sustains the rich heterogeneity, the potentially bountiful but ultimately vulnerable diversity of living assemblages. It is therefore critical to understand what it is that determines when boundaries are opened and when they are sealed between neighbours, and what the consequences are both for the participants themselves and for the systems they inhabit.

CHAPTER 7

ME AND YOU, US AND THEM: MERGER, TAKEOVER AND REJECTION

1. Encounters with Others—Threat and Promise

Loneliness is much more of a state of mind than a state of being. So great is the power of living beings to proliferate and diversify that, given the availability of water, a useable energy source, and moderate temperature, acidity and pressure, scarcely a nook or cranny of the earth's complex surface remains unpopulated for long. Encounters with other beings are among the most inevitable consequences of being alive. Yet both the scientific and emotional attitudes of human beings towards such encounters are curiously ambivalent. Why should this be?

Scientifically, the occurrence of more than one kind of organism—and even more than one component of an organism—in the same arena complicates the interpretation and predictivity of data. If the system is not "pure", it is difficult to be certain about the exact origin of observed phenomena and the distinction between cause and effect.

The purification of systems by means of eliminating "contaminants", has therefore been one of the most stringently pursued ideals in the development and application of analytical biological science. Freedom from contamination, the thinking goes, allows certainty and freedom from error.

The benefits and costs of the quest for purity are vividly illustrated by the study of microorganisms. There can be no doubt that the development of "aseptic" techniques enabling bacteria and fungi to be grown as "pure cultures" revolutionized understanding of these organisms and prepared the stage for the modern era of "biotechnology".

At first, pure culture techniques were of most value in diagnosing the roles of microorganisms as causes of disease. A famous set of maxims, "Koch's postulates", were formulated as criteria for the formal proof of these roles. First there had to be a constant association between the presence of the microorganism and the presence of the symptoms. Second, the organism was to be isolated from the affected organism or tissues and grown in pure culture. Third, the pure culture was to be used as a source of inoculum from which the microorganism could be artificially introduced into the healthy tissue or organism and cause the symptoms originally observed.

Fourthly the microrganism needed to be re-isolated and shown to be the same as the original one.

Pure culture techniques then became used increasingly as a means of revealing many aspects of the biology of microorganisms, notably their modes of proliferation and development and their responses to chemical and physical factors in their environment. Studies of the ability of cultures and culture extracts to bring about chemical reactions led to the elucidation of important enzyme-catalysed biochemical pathways. Several microorganisms, especially the bacterium *Escherichia coli*, the yeast, *Saccharomyces cerevisiae* and the mycelial fungus, *Neurospora crassa* became extraordinarily important "model organisms" in analytical studies of genes and their products. These studies contributed in no small way to the current understanding and application of molecular biology.

It would be very easy to get carried away by such success stories, and many researchers continue to re-inforce the purist line that has so clearly proved itself. However, there are two dangers in such reinforcement. The first, perhaps best highlighted by the discovery of penicillin following the chance contamination of a bacterial culture, is that it may impede the acquisition of important new knowledge or insights. More generally, it may distort understanding of the properties of organisms in a realistic environmental context where heterogeneity is the rule. One of the most fundamental uncertainties in biological science lies in applying results obtained from experiments made in simplified and often artificially contrived circumstances to the vastly more complex conditions in natural environments. The certainty with which events can be followed and interpreted in a well defined experimental system is inevitably accompanied by uncertainty about the relevance of the findings in a wider field.

Emotionally, both loneliness and company are viewed with a mixture of desire, dislike and fear. Loneliness is sought after for the sense of simplicity, isolation, freedom and self-awareness that it brings, but avoided because it also entails boredom, exposure, inadequacy and introspection. Company is sought after for interest, protection, support and self-absorption but avoided because it causes complications, exposes to threats, imposes constraints and implies loss of individual control.

Fundamentally human though all these emotional viewpoints may seem to be, their origin may be traceable to the risks and benefits that encounters with others can bring at all levels of biological organization. As outlined in Chapter 1, risks come in a variety of forms: restriction and wastage of resource supplies due to mutual competition; dysfunction due to transmission of disease-causing factors or mismatched components or activities; constrainment, and takeover by another entity. Benefits, on the other hand come from mutual protection, gaining access to resources within another's boundary and complementation between components or activities.

It is therefore not surprising to find widespread evidence for a counteractive interplay between "rejection" mechanisms which seal off and "access" mechanisms which open up boundaries between neighbours. I will now describe how these mechanisms come into effect at different hierarchical levels of organization before considering their significance during evolutionary differentiation and integration.

2. Rejection—Incompatibility, Self-protection and Self-defence

By "rejection", I mean the active or passive prevention of integration of contextual boundaries. It can take different forms, involve diverse mechanisms and have distinctive consequences.

In order to understand the varied forms and outcomes of rejection, it is necessary to define the dynamic contextual boundaries across or within which it occurs. However, this need is often not acknowledged, so that words like "resistance", "defence", "immunity", "antagonism" and "incompatibility" are used interchangeably and often emotively, leading to confused and sometimes heated arguments.

In the following discussions, active or passive mechanisms that involve keeping some distance or maintaining an intact boundary between entities will be referred to as "self-protection". These mechanisms commonly serve to isolate an entity from potentially harmful non-living environmental factors as well as from neighbours. Active mechanisms of self-protection involve "evasion" of hostile agents and "repair" of damaged boundaries. Repair is easily confused with "defence" (see below), but is directed primarily at boundary-sealing rather than arresting an incursion as such.

Truly active mechanisms of "self-defence" involve deploying resistive countermeasures specifically in the face of actual challenge. They imply some form of "recognition" of and "response" to an identifiable threat.

Self-protection and self-defence therefore occur at the actual interfaces between entities or systems (i.e. where these entities or systems are regarded as contexts in their own right).

"Incompatibility", on the other hand, describes the "interference" (due to mismatch or non-complementation) and/or competition (due to structural non-integration) which results from rejection between entities within the same contextual boundary. In some circumstances incompatibility can lead to the persistent subdivision of a system into separate subdomains; stability is thereby achieved by means of segregation. In other cases incompatibility results in instability; particular constituents are liable to overwhelm others or the system itself may degenerate due to internal conflict and the elimination of vital components.

The varied consequences of incompatibility may appear to be warlike and calculated to enhance the evolutionary fitness of the participants. Indeed, the very terms used to describe rejection may re-inforce this notion. However, it is important to realize that although the consequences of incompatibility can indeed be adaptive, they in fact follow automatically from the non-integration of reproducible entities within a common boundary. They are therefore organizationally impelled (see Chapter 4). In other words, "war" is a natural consequence of incompatibility, rather than an invention to suit an evolutionary purpose.

Particularly where indeterminate systems are involved, interpretation of the consequences of incompatibility is complicated by problems of boundary definition and differences in organizational scale between the participants. What may be defined as incompatibility within one context may lead to subdivision into separate contexts and so appear to play a role in self-defence. This is because the expression of incompatibility at any particular organizational scale depends on integrative processes at some larger scale in order that the entities concerned become enclosed within the same arena (cf. Chapter 1; Fig. 1.1).

For example, two genetically different mycelia of the same fungal species may "co-exist" indefinitely in adjacent but separate subdomains within a log because of interference between intracellular populations of nuclei and mitochondria (see below). This subcellular incompatibility is itself the consequence of the cellular "compatibility" that allows hyphal fusion. So, compatibility at one level of organization leads to interference at a lower level and to segregation at a higher level! Confused? Most of us are as long as we disregard the importance of defining contextual boundaries.

In the example just given, the participants are organized at the same scale, so that integration involves coalescence (Chapter 1). Where integration involves enclosure of a smaller "invader" by a larger "host", the issue arises as to whether the invader interacts compatibly (i.e. by integration) or incompatibly (by non-integration) with the components of the host. Incompatibility with host components may then lead either to the proliferation of the invader or to its segregation, elimination and/or consumption.

In discussing the consequences of incompatibility, it is therefore crucial to be clear about the context within which the rejecting entities are judged to be interacting. In the following discussions, interference or competition between entities within cellular boundaries will be referred to as "molecular incompatibility" whereas interference or competition between cells in multicellular assemblages will be termed "cellular incompatibility".

Finally, there is the delicate problem of deciding what should be regarded as "self". As was mentioned in Chapter 1, the concept of "self" has become inextricably linked with the concept of evolutionary survival units and hence with the possession of identical genes. However, a purely genetic definition of self is insufficient to cover all eventualities. Even genetically identical entities may lack mechanisms that allow them to integrate, whereas genetically disparate entities may be capable of integration. The only way out of this impasse is to acknowledge that there can be no realistic, absolute definition of self, just as there can be no realistic, absolute definition of individual.

Self can therefore only be defined relatively, in terms of whether discernible entities within some arena have recently proliferated from a common source *and* are capable of integrating. Only if *both* of these conditions are satisfied, can the entities belong to the same self. Rejection between entities, whether or not they are genetically identical, implies that the entities are not self. Equally, if two entities arrive within the same arena from genetically separate origins, then they are not self, even if they are or become capable of integrating. However, if such entities integrate, and remain integrated through subsequent generations, then they can be regarded as components of the same self, until or unless they dissociate.

2.1. Molecular Incompatibility

Molecular incompatibility, due to interference or competition between entities within cell boundaries, may be traced to the presence of nucleic acids from disparate sources. These nucleic acids may be packaged in mobile elements, plasmids, viruses, chromosomes, or nuclear, mitochondrial or chloroplast genomes.

Molecular competition results from the inability of separate informational units to integrate structurally, and so be subject to common controls governing replication and transcription processes. Disparities in these processes can then rapidly be amplified to the point where a particular kind of unit predominates within the cell.

Molecular interference can be due to synthesis of mismatched and therefore dysfunctional components of enzymes, organelles and other structures. Alternatively it can result from the production of substances which disrupt cellular functions or directly destroy other units.

As with all other forms of incompatibility, the consequence of molecular competition or interference is therefore either the elimination of one entity (by takeover or dilution) or persistent conflict. Elimination results in an obvious short-term evolutionary benefit to one of the entities, since it represents an extension or preservation of genetic domain or territory within which self-proliferation can occur, and an equivalent loss for the other entity. However, the benefit to the dominant entity is limited by the fact that it fails to gain information (as opposed to territory) from the other entity, and where dependent on the latter is ultimately "cutting off its nose to spite its face". Persistent conflict, on the other hand, may be detrimental to both entities but the degeneration that it leads to may serve to protect the boundaries of larger systems.

The potential for both elimination and persistent conflict can, by definition, be prevented by integration but, upon dis-integration will immediately be re-instated. As will be described, this situation, which first becomes apparent at molecular scales of interaction, recurs at all organizational scales.

Some of the clearest examples of molecular incompatibility are provided by plasmids and viruses. These examples have in their turn provided both serious problems and major opportunities for humankind in the development of genetic engineering technologies.

A single cell of the bacterium, *Escherichia coli*, can contain up to seven different kinds of plasmids. However, not all the kinds of plasmids capable of inhabiting *E. coli* can co-exist in this way. Rather they are classifiable into incompatibility groups. Whereas members of different groups can co-exist, where two members of the same group are present, one or other will quite rapidly be lost from the cell.

E. coli is also commonly infected by viruses, known as bacteriophages, whose nucleic acids show varied abilities to co-exist with the bacterial chromosome. As mentioned in Chapter 4, some phages, such as phage lambda, are known as "temperate". They can either undergo a "lytic cycle" in which they destroy the host cell or a "lysogenic cycle", during which their DNA is integrated and replicated with the host chromosome. It has been shown that maintenance of the lysogenic cycle is dependent on the production by the virus of a suppressor protein which prevents expression of those genes that lead to lysis. However, if this suppression is disrupted in some way, for example by DNA-damaging ultraviolet irradiation, then the lytic cycle will be initiated. The lytic cycle is therefore an expression of incompatibility which is suppressed when the virus DNA is integrated with host DNA and hence "maintained within bounds".

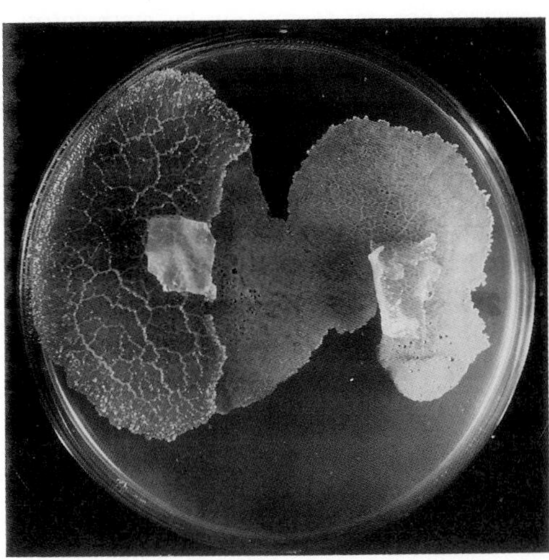

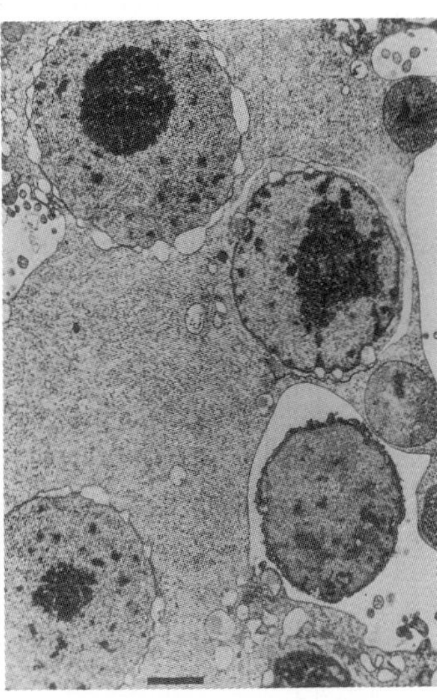

Figure 7.1. Lethal reaction in the slime mould *Physarum*. Left photograph, two incompatible plasmodia 17h after fusion—the "sensitive" strain on the right is dead, whereas the "killer" strain on the left is alive and will in due course invade the sensitive strain's territory. (From Carlile, 1972) Right photograph, electron micrograph showing two normal-looking killer strain nuclei on the left and two degenerating sensitive strain nuclei on the right (as well as two mitochondria on the far right). The upper sensitive nucleus has a disintegrating nucleolus and chromatin, and is being separated from the surrounding cytoplasm by a furrow. The lower sensitive nucleus has further disintegrated and has been expelled from the plasmodium. (From Lane and Carlile, 1979)

Whereas the relationships between temperate phages and their host genomes depend on the genetic content and expression of the virus, in other host-phage relationships the infectivity of the virus can be "restricted" or "modified" by the host. It has been found that when certain phages, e.g. phage lambda, are recovered from one strain, e.g. strain C, of *E. coli* and used to inoculate another strain, e.g. strain K, their infectivity in the latter is low (restricted) whilst remaining high in the source strain. By contrast, phages recovered from strain K have high infectivity if re-inoculated directly into strain K, but low infectivity if grown in strain C prior to re-inoculation into K. In other words, phage lambda can be restricted or modified, so possessing either low or high infectivity in K, depending on where it has been replicating prior to infection.

Such restriction and modification effects are due to the production, via the host genome, of "DNA-restriction" and "DNA-modification" enzymes. Restriction enzymes recognize and cut incoming DNA at sites determined by specific sequences of base pairs whilst the modifying enzymes prevent this action by methylating the bases at these same sites. The modifying enzymes protect the host from its own restriction enzymes, but may eventually also protect any incoming DNA that escapes restriction. Interestingly, the discovery of these restriction-modification mechanisms was initially regarded as a curiosity of no practical importance, but restriction enzymes now provide one of the most fundamental tools for the analysis and manipulation of DNA.

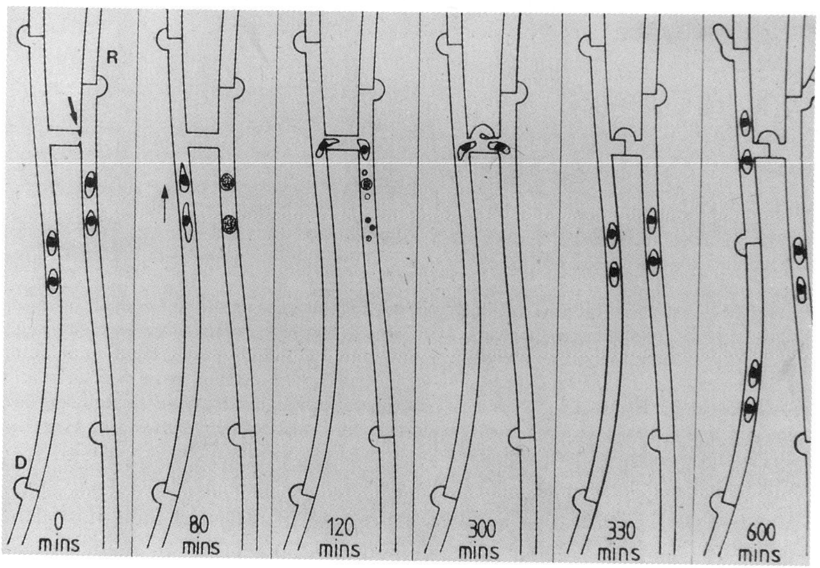

Figure 7.2. The nuclear replacement reaction in recipient hyphal compartments (R) following self-fusion with donor hyphae (D) of dikaryotic basidiomycetes containing two nuclei per compartment. (From Aylmore and Todd, 1984)

The destruction of one genome in the presence of another has also been found to occur in the eukaryotic cellular systems of fungi and slime moulds. However, the underlying mechanisms are not as clearly understood as in the restriction-modification systems. In some cases they allow takeover of protoplasmic domain. In slime moulds, takeover follows passage of a zone of lysis across susceptible plasmodia (Fig 7.1).

As was mentioned in Chapter 4, septate (internally partitioned or compartmented) hyphae produced by the same or closely related species of many fungi have the ability to fuse or anastomose. In some cases, a remarkable "nuclear replacement reaction" occurs within the participant hyphae, even if they contain genetically identical nuclei and protoplasm, and indeed even if they arise from the same mycelium. During this reaction, one hypha behaves as a "donor" and the other as a "recipient", such that the nuclei or nucleus in the recipient compartment disintegrate(s) and become(s) replaced by the daughter(s) of the nuclei or nucleus in the donor (Fig. 7.2).

The reason for this apparent nuclear civil war in self-fusions is still unclear, although it seems to be a feature of those fungi which have hyphal compartments containing just one or two nuclei. A possible explanation is that it is necessary not only for the nuclei to contain the same genetic information if they are to co-exist stably, but also that they should be expressing this information equally or complementarily. Other examples where the number of nuclei in common protoplasm is reduced to some specific value are found amongst eukaryotes, notably in the production of "egg nuclei" following meiosis (see Chapter 4). However, in these cases the nuclei are not genetically identical.

The fusion of hyphae containing nuclei of different genetic origin can also lead to nuclear replacement, and indeed wholesale nuclear invasion from one mycelium into another (Fig. 7.3), paralleling the situation described above for slime moulds. More generally, the presence of incompatible nuclei within the same protoplasm leads to a mutual rejection response, such that the protoplasm degenerates, causing a "demarcation zone" to develop between paired mycelia (Figs 7.4, 7.5). This demarcation zone is often coloured by the production of the dark pigment, melanin, due to the action of phenol-oxidizing enzymes. Similar mutual rejection can occur between slime mould plasmodia (Fig. 7.6). The resulting failure to integrate protoplasm has been referred to as "somatic incompatibility" in order to distinguish it from incompatibility expressed between specialized sex cells (see below).

The cause of somatic incompatibility in fungi is not fully understood. However, one potentially plausible explanation is that co-expression of disparate genetic information leads to the dysfunction of pathways that transfer electrons to oxygen, for example within mitochondria. This leads in turn to the production of reactive oxygen species, the activation of phenol-generating pathways and phenol-oxidizing enzymes, and the generation of highly reactive free radicals. The latter disrupt cellular structure and functioning, generate melanin and allow the release of protein- and nucleic acid-destroying enzymes which cause the protoplasm to digest itself

(Fig. 7.7). Such autodigestion strongly resembles that exhibited during such degenerative processes as protoplasmic "senescence" and apoptosis or "programmed cell death" (see Chapter 5, regarding mechanisms of cellular degeneration).

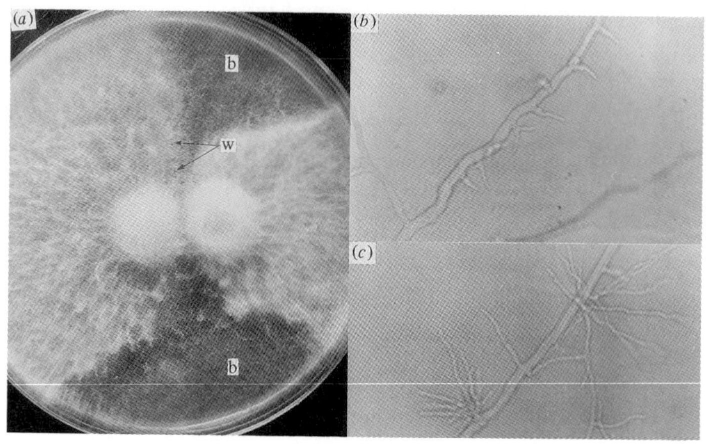

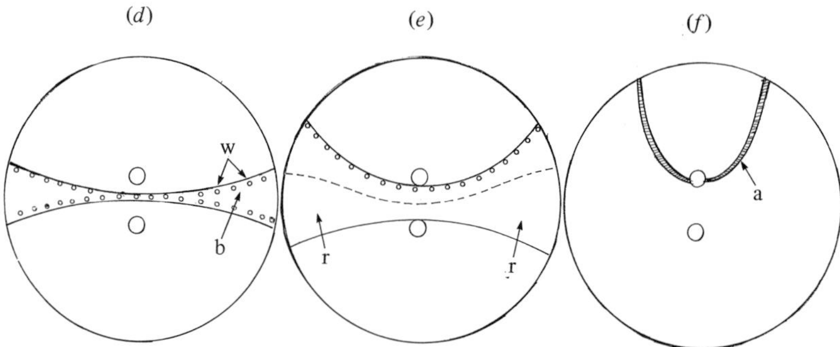

Figure 7.3. Nuclear invasion following fusion between incompatible mycelial networks of the basidiomycete fungus, *Stereum hirsutum*. (*a*) An asymmetric, bow-tie-shaped region, b, of degenerating mycelium, bounded by regions where watery droplets, w, have been exuded has spread from the interface between two mycelia which have been paired in the centre of a Petri dish. (*b,c*) Unusual branching patterns of hyphae in the degenerative region. (*d-f*) Typical stages in the development of the reaction, culminating in the extension, following regeneration, r, of one network's territory at the expense of the other and formation of a persistent zone of mutual antagonism, a. (From Coates et al., 1981).

Whatever the chain of events that leads to expression of somatic incompatibility, it has profound consequences because it prevents physiological integration between genetically different mycelia in natural populations. Also, since genetic control of the rejection response is generally complex, involving many genes and probably many different forms (alleles) of the same genetic loci, virtually any two unrelated mycelia, and often even related mycelia, will be incompatible. Natural populations of many fungal species therefore consist of innumerable individual mycelia, each of which is genetically homogeneous within its own boundaries and distinct from others. The mycelial systems of fungi therefore exhibit, at the cellular level, territorial patterns whose genetic control, expression and evolutionary consequences are echoed in all kinds of socially and developmentally indeterminate systems.

Figure 7.4. Mutual rejection (somatic incompatibility) leading to the formation of a demarcation zone in the midline between two genetically different mycelia of the same fungal species (*Coriolus versicolor*) cultured opposite one another in a Petri dish. (From Rayner and Todd, 1977). This zone corresponds with the formation of interactive boundaries between decay columns in wood (see Figure 2.21).

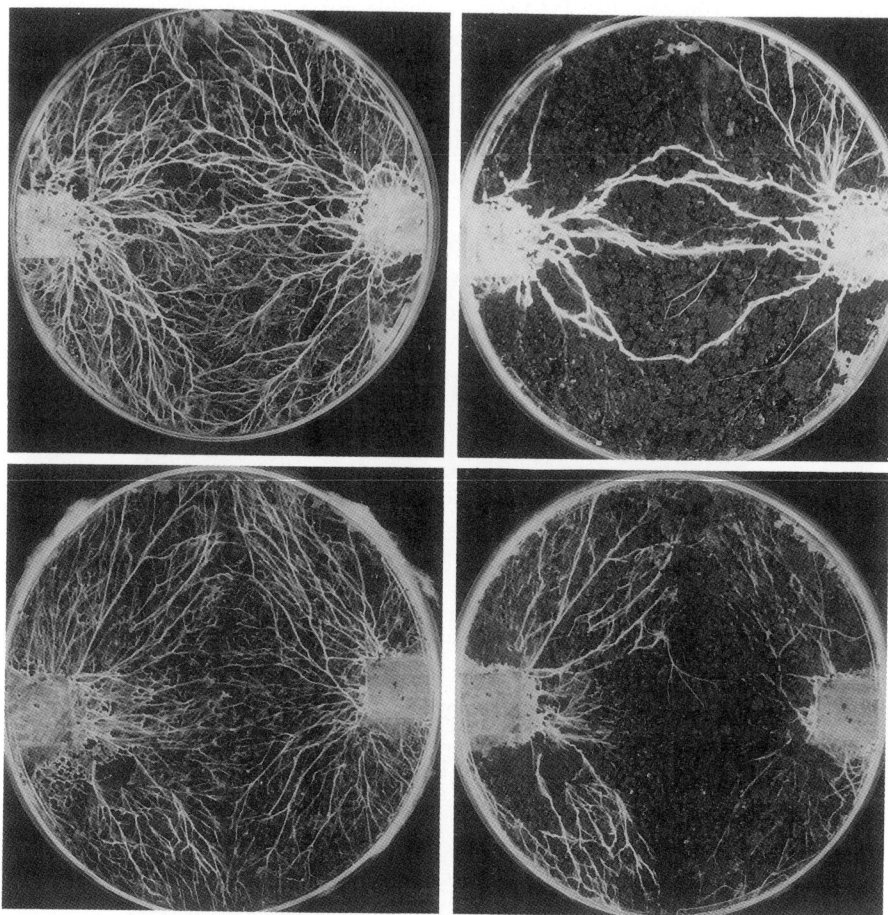

Figure 7.5. Consequences of anastomosis between mycelial cord systems of the basidiomycete fungus, *Phanerochaete velutina*, growing out from beechwood blocks inoculated into 14 cm diameter Petri dishes containing unsterilized soil. The top two photographs show early (left) and late stages of interaction leading to the reinforcement of a few cords connecting woodblocks occupied by genetically identical mycelia. The lower two photographs show the development of an unoccupied "no man's land" between persistent, mutually incompatible, genetically unidentical mycelia. (From Dowson et al., 1988b).

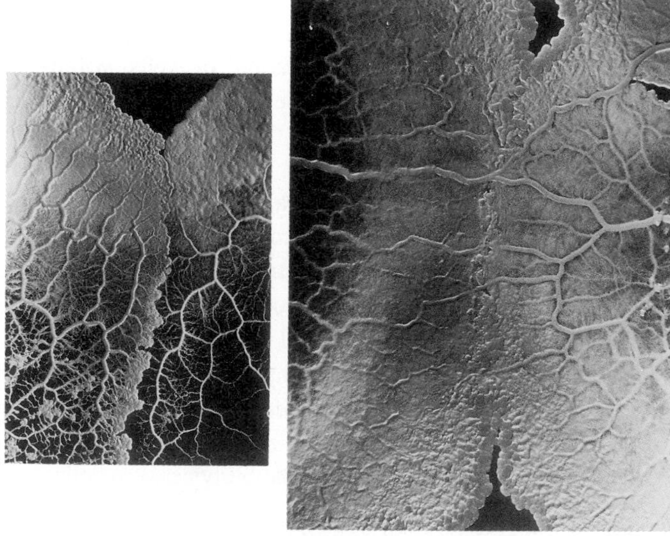

Figure 7.6. Rejection (left) and integration (right) between plasmodia of the slime mould, *Physarum polycephalum*. (Reprinted with permission from *Nature* (Carlile and Dee, 1967), copyright (1967) Macmillan Magazines Ltd)

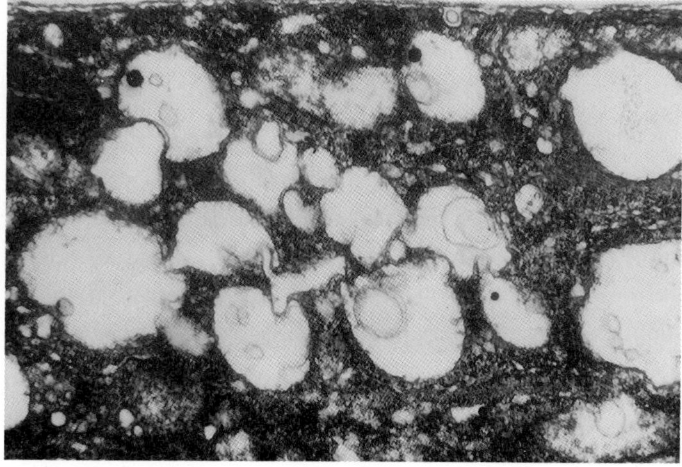

Figure 7.7. Protoplasmic degeneration, as seen under the electron microscope, due to somatic incompatibility in the fungus, *Phanerochaete velutina*. (From Aylmore and Todd, 1986, by permission of the Society for General Microbiology).

2.2. Cellular Incompatibility

Cellular incompatibility, the non-integration of cells within multicellular assemblages, may partly be traced to the absence of such mechanisms as hyphal fusions, gap junctions and plasmodesmata, that open communication channels between neighbours (see Chapter 4). It can allow a particular type of cell to proliferate more rapidly than others and to kill and draw resources from them, resulting in various kinds of intercellular replacement and parasitism.

Some "killer cells" release toxins that destroy their neighbours. Two examples are "killer" yeasts and "killer" *Paramecium* (a ciliate protist—see Chapter 2). These cells contain nucleic acid of possibly viral origin in addition to their own chromosomes. This additional nucleic acid consists of double-stranded RNA in the yeasts and "kappa particles" in the *Paramecium*. These examples provide an interesting case where compatibility between disparate molecular units (host and viral genomes) within a cell can lead to incompatibility between cells.

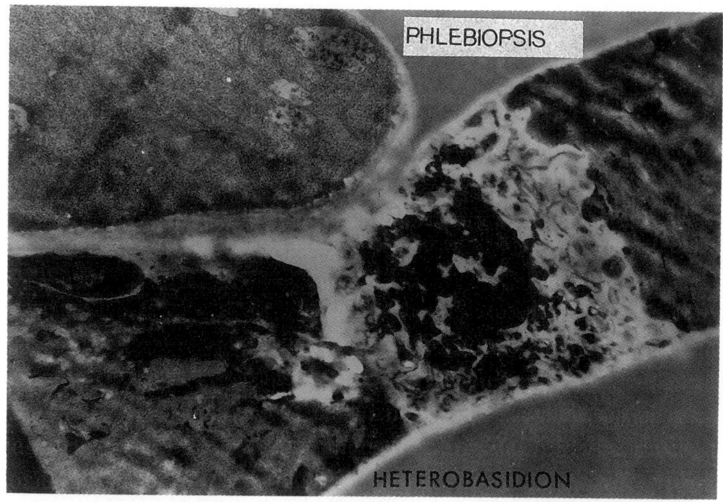

Figure 7.8. Electron micrograph showing hyphal interference and resultant degeneration of protoplasm in a hypha of the basidiomycete fungus, *Heterobasidion annosum*, following contact with a hypha of *Phlebiopsis gigantea*. (Photograph by F.E.O. Ikediugwu)

Amongst mycelial fungi, two distinctive phenomena, known as "hyphal interference" and "mycoparasitism" commonly follow close encounters between non-anastomosing hyphae. Hyphal interference typically occurs between different species and results in rapid protoplasmic death within the affected hyphal compartments of one or both participants (Fig. 7.8). The process shows a strong resemblance to somatic incompatibility reactions, and like the latter seems likely to involve the same cascade of metabolic processes leading to the generation of reactive oxygen species, phenolic compounds and free radicals. In some cases, mutual hyphal interference may be localized to the interface between interacting mycelial systems, resulting in a stalemate. In other cases, unilateral interference reactions encroach into one of the systems as it becomes invaded and replaced by the other system. Mycoparasitism involves the encoiling or indeed the penetration of hyphae by other hyphae and appears to depend on the absence of a hyphal interference reaction. It can occur within individual systems, giving rise to "self-parasitism" (Fig. 7.9), as well as between individuals, both of the same and of different species.

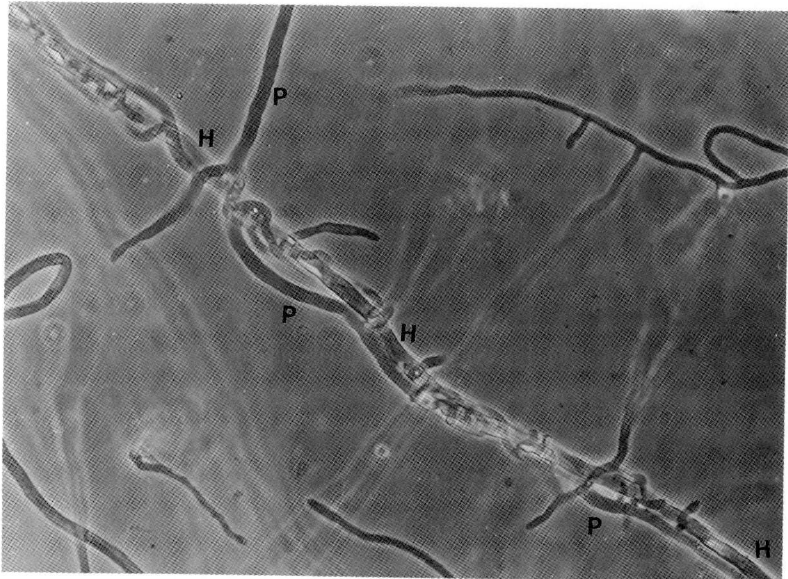

Figure 7.9. Self-parasitism, resulting in the encoiling of a "host" hypha (H) of the basidiomycete fungus, *Stereum hirsutum*, by other, "parasitic" hyphae (P) from the same mycelium. (Courtesy of A.M. Ainsworth).

Reactions analogous to hyphal interference occur when plant cells undergo what is known as a "hypersensitive" reaction in response to close encounters with fungal cells. This reaction again involves rapid protoplasmic death, and again seems likely to involve generation of phenolic compounds, reactive oxygen and free radicals. It prevents the establishment of fungal parasites (known as "biotrophs") which depend on being able to draw resources from living plant tissues. It is therefore often regarded as a "resistance" mechanism, a kind of "suicidal self-defence" which seals off the remainder of the plant tissues from attack.

The resistance of plants to infection by fungi which induce a hypersensitive reaction is often attributed to the toxicity to fungi of compounds, known as "phytoalexins", produced by the dying cells. However, whether these compounds are actually or incidentally self-defensive is debatable because the dead cells themselves may provide an effective self-protective barrier (see below). Production of phenolics and similar compounds allows boundary-sealing in plants, just as it does in fungi, and can be induced by physical or chemical damage as well as the presence of another organism. Also, there are many fungi (known as "nectrotrophs") which kill plant cells in advance of drawing resources from them (see below). As with interfungal interactions, death of protoplasm is therefore an expression of incompatibility which can both hinder and herald the advance of an invader.

Another example of induced cell death is found in the immune system of vertebrates, where host cells that have become infected by viruses are destroyed by cytotoxic T cells (see Chapter 4). The infected host cells are identified by the presence of peptide fragments derived from the virus which are bound to "MHC-proteins" (see below) and delivered and displayed by these proteins on the surface of the cell. The MHC-peptide complex acts as a ligand which binds to T cell receptors and co-receptors. The infected cell is then killed by a mechanism which is still not certain, but probably involves apoptosis, rendering the cell unavailable for proliferation of the virus.

In this case, the host cell becomes "labelled" by MHC-peptide complexes on its surface and is commonly said to be "recognized" by the immune system. Other parts of the immune system also work by labelling invasive cells or molecules as nonself, heralding "attack" by phagocytes and a system of proteins known as "complement". In insects, there is also evidence that invasive cells become labelled for engulfment via the action of phenol-oxidizing enzymes. In all these cases, some means of recognizing and marking cells as nonself is therefore prerequisite to self-defence. However, in understanding the origins of these systems it may not be wise to think only about those multicellular organisms in which they have attained their greatest sophistication.

2.3. Tissue Rejection

The advent of transplant surgery has made many people aware of the phenomenon of graft or tissue rejection—the non-integration of cell assemblages between donor and recipient human beings. However, the widespread occurrence of this phenomenon amongst all kinds of animals and (to a lesser extent) plants, as well as its analogy with the somatic incompatibility systems of such organisms as fungi and slime moulds, is less generally appreciated.

Amongst determinate animals, tissue rejection is normally observed only under artificial circumstances, when grafts are made. However there are some animals in which tissue rejection is a regular and important feature of their natural lives, just as somatic incompatibility is amongst fungi and slime moulds. These animals are the colonial invertebrates first introduced in Chapter 2; the colonial sponges, cnidarians, sea squirts and bryozoans.

Work with colonial invertebrates has shown that in natural populations closely related or genetically identical individuals are able to fuse, integrating their tissues to the extent of forming common digestive and vascular systems. Genetically disparate individuals, on the other hand reject, either segregating into separate territories or invading one another destructively.

There is also evidence, as in fungi and slime moulds, that the genetic systems underlying invertebrate rejection are highly diverse ("polymorphic") and that the intensity of the response varies with the degree of genetic disparity. The polymorphism is due to the involvement of multiple loci and/or multiple alleles, and divides populations into innumerable individual lines. The variation in intensity may indicate the involvement of a few genes with major effects and many genes with minor but additively important effects. However, only in one case, the colonial sea-squirt, *Botryllus schlosseri*, is genetic control of tissue rejection reasonably well characterized. Here, in order for rejection to occur, two diploid individuals must have no common alleles at a single highly polymorphic locus. If two individuals share one or both alleles, they can fuse. As many as around 100 alleles of this locus are thought to occur.

Studies of graft rejection in mammals led to the discovery of both major and minor "histocompatibility" genes, which need to be different to invoke a rejection response. The major histocompatibility genes are clustered together as a complex ("MHC") on the same chromosome and code for the MHC proteins (see above). These proteins were first characterized in mice, where they are referred to as "H-2 antigens" (histocompatibility-2 antigens). In human beings they are called "HLA antigens" (human leucocyte-associated antigens) because they were first demonstrated on white blood cells (leucocytes).

Not only does every individual possess five or more MHC genes, but there can be large numbers of alleles (as many as 100) at each locus. It is therefore very rare to find two individuals with identical MHC proteins in natural populations. This makes it difficult to match donor and recipient for organ transplant operations and parallels the situation in populations of colonial invertebrates, fungi and slime moulds.

As mentioned earlier, the recognition of MHC proteins as nonself by T cell receptors brings the cell-mediated component of the vertebrate immune system into operation. Normally, for this to happen it is necessary for the MHC proteins to be associated with foreign peptide derived from infective microorganisms or viruses. However, the presence of nonself MHC proteins on introduced tissue will likewise trigger immune rejection.

Transplant rejection is therefore commonly thought to be an incidental by-product of the way in which vertebrate bodies recognize and defend themselves against infections. The fact that MHC systems are so very polymorphic has correspondingly been attributed to the huge variety of different kinds of infections that an organism may be exposed to, and the need to limit the spread of infections between neighbours. The latter idea can be understood in terms of the fact that if one organism proved susceptible to an infection, then all its neighbours would be equally susceptible if they had identical MHC molecules.

However, this way of thinking may have been biased by focusing on the functioning of MHC systems only in a particular set of organisms. In view of the widespread natural occurrence of tissue rejection amongst colonial organisms, including the sea squirts (from which the vertebrate line probably arose by neoteny; Chapter 5), it might be better to think of immune defence against infection as an incidental by-product of incompatibility between neighbours.

Also, although organs or tissues don't normally travel from one vertebrate to another, odours certainly do, and there has been some remarkable work with mice and rats which suggests that these animals can actually smell the difference between MHC molecules or derivatives, especially in urine. There is therefore a real possibility that these molecules can allow recognition and response at the behavioural level, between separate animals. Since the number of encounters with others during an individual's life span may be very large, then a correspondingly large number of MHC specificities would be required to detect nonself differences.

2.4. Shields, Avoidance and Noxiousness

All mechanisms which isolate whole organisms from one another result in incompatibility at population and community levels of organization. Some of these mechanisms are due to incompatibilities expressed at molecular, cellular or tissue levels of organization. Other mechanisms are due to self-protection and self-defence operating specifically at what are, or what amount to, the external boundaries of the organisms themselves. These mechanisms will be considered in this and the next section.

Passive self-protective mechanisms may generally be classified as physical, chemical or cryptic. Physically insulating mechanisms make use of the external boundary either of the organism itself or of the context that it inhabits, as a protective shield or shelter. Chemical mechanisms are based on production of noxious substances. Cryptic mechanisms of self-protection involve avoidance of detection by an adversary.

Physically insulating boundaries may consist of components derived, directly or indirectly, from within the organisms themselves, e.g. cell walls, exoskeletons, rinds and shells. Alternatively, it may incorporate extraneous materials. These may either be actively assembled from the environment, as in the casings of certain polychaete worms, clothing, birds' nests, termite mounds and other buildings, or be present in pre-existing refuges such as empty shells, buildings or cavities in rocks.

All these mechanical devices enable organisms to reduce losses to outside agents of all kinds, both living and non-living, and so to conserve energy. Various processes such as thickening, hardening, impermeabilization and outgrowth of down increase the insulating effects of the cover, whilst spiky protrusions generally protect against ingestion by other organisms. However, effective conservation by means of effective insulation implies limited ability to exploit conditions of plenty and may also prevent or retard elimination of harmful agents from within the body (see Chapter 6).

Any breach in the insulation requires urgent repair if it is not to result in loss of function and/or resources. Often, repair is not fast enough to prevent some loss of function, and so all that can be achieved is some form of damage-limitation involving the assembly of a new boundary inside the previous one. The organism's frontier then appears to have retreated within itself. Since other organisms may invade the abandoned regions, but not progress beyond this new frontier, processes producing the latter are commonly regarded as defensive. However, it may be better to regard them as self-protective, i.e. the consequence of maintenance of an external boundary.

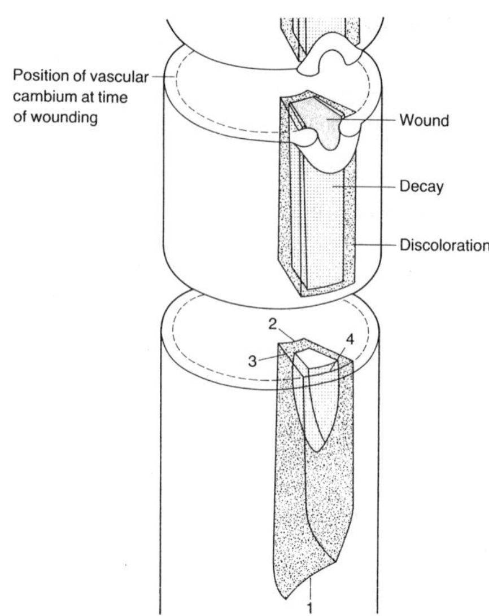

Figure 7.10. Diagram illustrating typical patterns of decay and discolouration originating from a deep wound, and location of putative "compartmentalizing walls 1, 2, 3 and 4" in a hardwood tree trunk. (From Cooke and Rayner, 1984)

A particular case where it is important to distinguish between defence and repair or maintenance concerns the way fungi causing decay and discolouration become established in the wood of trees following damage to or death of bark (see also Chapter 6). Like other organisms with indeterminate development, a tree can be thought of as a hydrodynamic or "plumbing" system. It connects sites of active uptake of water and mineral nutrients from soil with energy-gathering sites in a distant, photosynthesizing canopy. Between these sites are annually renewed axial and radial series of conduits in xylem, phloem and medullary rays, bounded on the outside by an insulating, water-resistant layer of corky ("suberized") tissue, bark.

If bark is killed or damaged, so that the seal it provides is broken, the underlying conducting sapwood effectively becomes exposed to the atmosphere and hence to dysfunction as a result of entry of air and drying. The degree to which dysfunction occurs depends on circumstances both within and outside the tree. For example, it will depend on whether water columns are under pressure (as they usually are in spring) or under tension (as they are in summer and autumn)—so allowing air to be sucked in when they are broken. It will also depend on the amount and depth of the initial damage, where it is located on the tree and the availability of water to the tree.

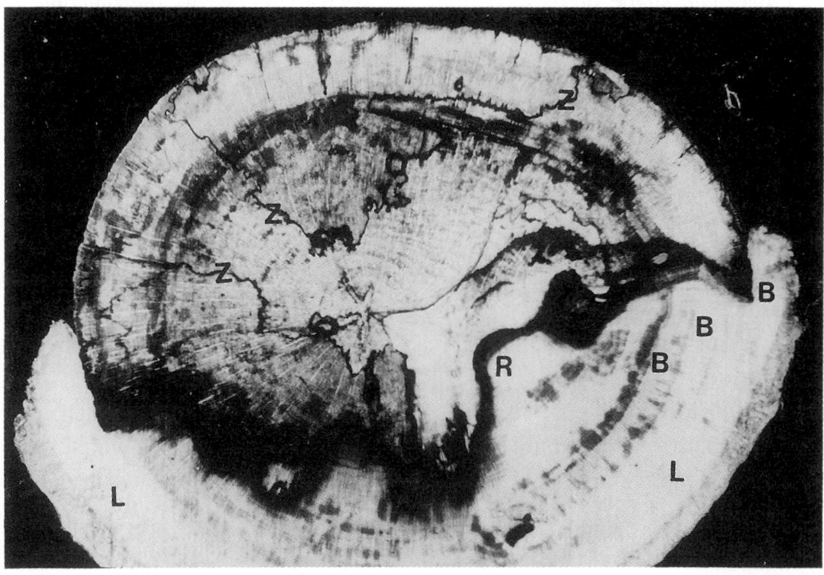

Figure 7.11. "Interactive plumbing" revealed in a cross section through a partially living, partially decaying branch of beech (*Fagus sylvatica*). Successive episodes of damage to the protective layer of bark, and resultant aeration and dysfunction of the wood cylinder, have resulted in the production of a concentric series of "barrier" zones (B), between wood present at and formed after the time of damage. Currently only about half the branch circumference is protected by an intact layer of bark and occupied by living wood (L). The unprotected remainder is dysfunctional, separated from the living wood by a dark "reaction zone" and colonized by mycelia of decay fungi. Territorial boundary zones (Z) between regions occupied by different mycelia either pass straight across or are deflected around the barrier zones. (From Rayner and Boddy, 1988; photograph by Ruth Wolstenholme)

The spread of dysfunction from the site of damage is eventually curtailed at locations where boundary-sealing processes take place. These processes involve the plugging of conduits with gums and ingrowths known as "tyloses", the death of living parenchyma cells and the associated impregnation of the tissues with phenolic substances. The latter may darken in the presence of air and phenol-oxidizing enzymes, so that a "reaction zone" appears at the interface between functional and dysfunctional tissues—which by this time are already likely to have been invaded by fungal hyphae. Wood subsequently formed by renewed cambial activity is also protected from the spread of dysfunction by means of a "barrier zone", an annual ring whose innermost cells are impregnated with suberin.

Depending on circumstances, the dysfunctional tissues may be confined to peripheral wood or take the form of axially elongated, often wedge-shaped columns

(Fig. 7.10). The shape of these regions has been explained as the outcome of an active defence mechanism, "compartmentalization" which produces a series of barriers or "walls" that arrest the incursion of fungi. The walls are said to increase progressively in strength, in the order 1 to 4, and correspond with the reaction zones and barrier zones which are in turn envisaged to be sites of a continuing dynamic conflict between fungal aggressor and tree defender. However, reaction zones and barrier zones are often produced in discontinuous series (Fig. 7.11) rather than being edged continuously backwards, and they are easily breached by fungi if the wood behind them becomes dysfunctional. When it is recalled that the restricted aeration in functional sapwood inhibits exploitative growth of mycelia (Chapter 6), it therefore seems more logical to regard reaction and barrier zones as reparative rather than defensive.

One of the reasons that reaction zones have been so readily interpreted as defensive is that they commonly (but not always) contain chemicals which are toxic to fungi. The adaptive argument was therefore advanced that trees would not spend the energy required to synthesize toxic compounds if they were of no use to them. However, this argument overlooks the fact that many hydrophobic compounds produced in response to oxidative stress by dying cells are intrinsically liable to be toxic—not only to fungi, but also to other organisms, including plants (cf. Chapter 5).

The above argument introduces the fact that production of substances that are toxic, or even just "unpleasant", can provide self-protection. There are numerous examples of plants and animals that are not generally eaten or parasitized because of their noxiousness to all but a few suitably specialized organisms. A classical example is the monarch butterfly, *Danaus plexippus*, which is avoided by birds because it contains "cardiac glycosides" that are toxic to vertebrates. These glycosides are derived from toxic milkweeds which monarch caterpillars are able to consume.

Toxic organisms often seem to "advertise" their presence by being strikingly coloured and hence highly conspicuous—a feature which can only be advantageous if the organisms that might consume them are capable of learning from a nasty experience. Moreover, many non-toxic organisms "mimic" the colouration of toxic ones and so are avoided. An example is the edible viceroy butterfly, *Limenitis archippus*, which is avoided by birds due to its resemblance to the monarch butterfly.

In all the examples just given, the toxic compounds are contained within the organisms and so protect against potential predators and parasites. In other cases toxins are actually released into the environment immediately surrounding an organism, and so may have the additional effect of reducing the incidence of competitive neighbours. Amongst microorganisms such released toxins are often referred to as antibiotics. Amongst plants, a well known case is the walnut tree, which sheds a toxic rain to the ground below its canopy. This kind of phenomenon is sometimes referred to as "allelopathy".

A familiar example of cryptic mechanisms of self-protection is visual camouflage. This can take a variety of forms, all but one of which depends on preventing detection of the location of the organism's external boundary. The exception is provided by mimicry—when the animal resembles a structure such as a leaf or a twig. The other examples include: being the same colour as a homogeneous background; being striped, mottled or spotted against a heterogeneous background; disguising the presence of an eye; having countershading so that one part of the organism stands out against a light or a dark background, so making it seem smaller and to have a different shape than it really has.

Active avoidance mechanisms can also take a variety of forms. These include moving or growing away from a source of danger, playing dead and undergoing changes of colouration. Colour changes may either provide camouflage against different backgrounds, as in chamaeleons, or, as in some cephalopods (octopuses, squids and cuttlefish) occur in such rapid sequence as to make the organism's boundary seem even more dynamic than it really is.

2.5. Weapons and Smokescreens

True active defence, the deployment of resistive countermeasures in the face of actual challenge, involves either the infliction of damage on an assailant or the placement of obstacles in its path. A well-known example of the latter is the production of a jet of inky fluid by cephalopods, which acts as a smokescreen behind which they can effect a retreat.

Many animals use parts or secretions of their body as weapons, biting or stinging adversaries or emitting toxic sprays or offensive odours. There are many familiar examples, but one of the most dramatic is provided by bombardier beetles. The beetles have a gland made up of two chambers in a rotatable turret which can be aimed at a predator. The chambers contain phenols, hydrogen peroxide, catalase and peroxidase enzymes which when mixed form a boiling, explosive cocktail containing toxic quinones which is fired out of the turret under pressure.

2.6. *Territoriality*

A territory is a region of physical space within which the life-sustaining processes of an organism or society are temporarily or persistently contained. The boundary of a territory is characteristically defined at interfaces with neighbouring territories where rejection occurs. When allowed to form naturally, it is generally irregular and possesses fractal properties at least over some range of scales. It can be likened to the watershed between river drainage basins.

The inhabitant of a territory can be spatially indeterminate, as with fungal mycelia and other "colonial" forms. Alternatively, it has a determinate body form but maps out an indeterminate "trajectory" as it patrols the territory over time, as with many animals.

The concept of territoriality was first developed as an explanation of the aggressive behaviour of vertebrate animals, especially mammals and birds. These animals characteristically attack intruders, the more vehemently so the closer they are to the centre of their territory. At the boundaries of their territories they exhibit "conflict behaviour", often characterized by threat displays if they are within the boundary and escape behaviour if they are not. These behaviours often appear to be ritualized, and may help to prevent the costs that might otherwise be incurred by infliction of serious injuries. The use of vocal calls and scent-spraying is a common means of marking territorial boundaries which otherwise might only be detected when an energy-wasting aggressive response is provoked.

In some cases territorial behaviour appears to have a controlling influence on population numbers. Here, territory size is maintained above a minimum level sufficient to support its occupant(s); if population numbers increase above carrying capacity (see Chapter 4), then some members of the population become marginalized and fail to survive.

There are obvious parallels between the territories of birds and mammals and those formed both at larger scales by human societies divided up by anything from parish to national boundaries, and those formed at smaller scales by colonial invertebrates, social insect societies and fungal mycelia.

3. Overriding Rejection—Gaining Acceptance

Being rejected, whether at cellular, multicellular or territorial interfaces is profoundly self-limiting. It puts the resources and aptitudes of a neighbour out of reach and so restricts opportunities for self-furtherment. Any mechanism which allows an entity to obviate rejection barriers will therefore enhance its prospects. There are two main ways in which this can be achieved—one subtle, the other brutish. In this section, the more subtle mechanisms will be considered.

3.1. Sexual Attraction, Courtship and Mating

Sexual union requires a high degree of intimacy, no matter how separate the participants may be for the remainder of their lives! Furthermore, if such intimacy is to be productive, it is necessary in many organisms for the participants to be genetically different at specific genes determining gender or "mating-type" (see below). In such cases, and especially where no other means of reproduction (e.g. by some form of clonal proliferation) is possible, evolutionary survival depends on integration between entities that are genetically not self. Other-than-self-interest then becomes of paramount self-interest!

For those people who might view evolutionary furtherance as the consequence of competitive struggle between rampantly individualistic entities, sex can therefore seem paradoxical if not bizarre. This difficulty may stem both from not accepting the importance of integrative processes in evolutionary progression and not seeing how co-operation can emerge from what may initially be extreme individualism. As will be emphasized later, there are cases where sexual acceptance can lead to a kind of piracy or parasitic takeover, in which the genes from only one rather than both participants are passed on to subsequent generations. This raises the possibility that sexual union could have originated as the outcome of a "bid" for genetic territory, that became "tamed" into merger.

Whatever its origins and consequences, if sexual union is to occur at all, rejection between nearself entities (see Chapter 1) needs to be prevented or circumvented in some way.

Some interesting insights into how this may be achieved come from observations of fungi. As has been mentioned, the hyphae of different individual mycelia of the same species of many fungi can fuse with one another so that their protoplasmic contents become integrated. In most of the large group of fungi known as "basidiomycetes"—which includes the familiar mushrooms, toadstools, brackets etc—sexual outcrossing depends on such fusion. Here, as shown in Fig. 7.12, nuclei and mitochondria are mixed at the fusion sites between neighbouring mycelial networks, so producing "heterokaryotic" hyphal compartments that contain physically separate but genetically different types of nuclei. Heterokaryotic hyphae may then emerge directly from the fusion sites, or a remarkable process of nuclear migration follows, during which nuclei from one or both mycelia invade the other in one or both directions. The invading nuclei proliferate and convert the recipient mycelium wholly or partially into a heterokaryon that contains populations of resident and immigrant nuclei.

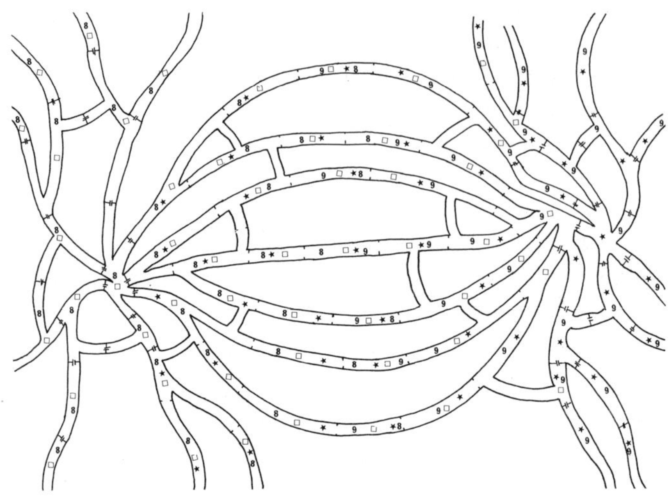

Figure 7.12. Diagram illustrating patterns of association and exchange of nuclei (represented as squares and stars) and mitochondria (represented as numbers) following fusion between sexually compatible homokaryotic mycelial networks of basidiomycete fungi. Septa are depicted as intact or eroded valve-like partitions (see Figure 7.13).

Like traffic negotiating a city centre, the migrating nuclei may follow one or other or both of two paths through a resident mycelium. On the one hand they may follow radial thoroughfares towards the origin of the resident network, in which case they are impeded by septal partitions that have to be eroded or by-passed to let them through (Fig. 7.13). On the other hand they follow the "ring roads" produced by lateral anastomoses. Whichever route is taken, mitochondria do not usually migrate with the nuclei and so only mix within the vicinity of the interface between mycelial systems.

In some cases, the heterokaryon that emerges from this nuclear exchange has two nuclei per compartment and so is known as a "dikaryon". Dikaryons commonly produce curious recurved hyphal fusions, known as "clamp connections" to one side of septal partitions.

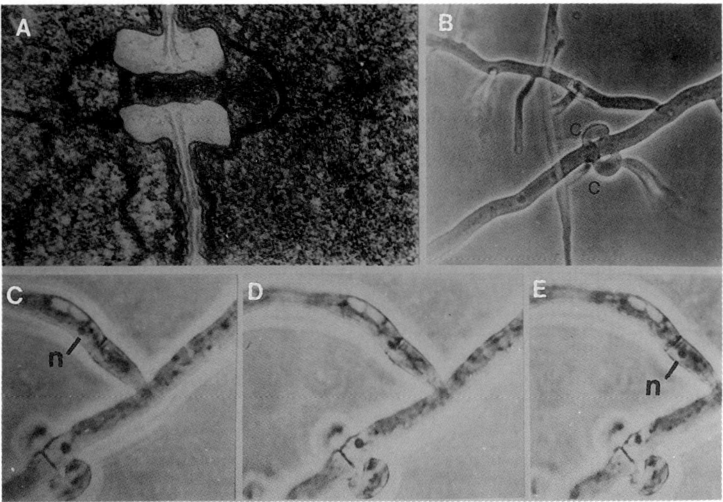

Figure 7.13. Nuclear migration and septal erosion in basidiomycete fungi. A, longitudinal section through a septum, viewed under the electron microscope, revealing its elaborate, valve-like structure. (Courtesy of R.C. Aylmore). B, a hypha of *Stereum hirsutum* which has undergone nuclear migration—septa normally present in the "clamp connections" (c) have been eroded. (Courtesy of A.M. Ainsworth). C-E, passage of a nucleus (n) through an eroded septum following mating in *S. hirsutum*. (From Ainsworth, 1987).

Once established, the heterokaryotic mycelia grow independently and are not accessed by further immigrant nuclei. However, they can fuse with, and their nuclei can invade as yet unmated "homokaryotic" mycelia, containing only one kind of nucleus. If, on the other hand, heterokaryons encounter and fuse with other heterokaryons, then a mutual rejection reaction occurs at their boundaries due to a degenerative somatic incompatibility reaction (see above). Eventually, once a heterokaryon has grown and assimilated sufficient resources, parts of it produce local hyphal proliferations which develop into "primordia", the early stages of fruit bodies. The fruit bodies enlarge and mature and only once they have started to produce specialized hyphal tips known as "basidia" do the two kinds of nuclei within the hyphae normally fuse (i.e. undergo "karyogamy") to form a single diploid nucleus. This nucleus then divides meiotically (see Chapter 4) to produce haploid daughter nuclei which become enclosed within groups of "basidiospores" borne on the outside of the basidium. The spores are dispersed from the fruit bodies and if they arrive at a suitable place germinate into "homokaryotic" mycelia which become divided into multinucleate or uninucleate compartments. In the latter case, the mycelia are "monokaryons".

There are several fascinating and unusual features in basidiomycete mating that provide some important insights into the kind of relationships that can develop between disparate nuclear and mitochondrial genomes inhabiting the same protoplasm.

Firstly, the initial stage of sexual union, the integration of protoplasm ("plasmogamy") occurs between "somatic" (i.e. sexually unspecialized) hyphae. However, the usual consequence of fusion between such structures would ordinarily be expected to be rejection, due to somatic incompatibility. That this expectation is justified is clear because heterokaryons reject one another, even if they have one nuclear type in common and their non-alike nuclei would be compatible if contained in homokaryons. Also, even homokaryons that can mate because they have complementary mating-type genes (see later) commonly show initial signs of rejection before the heterokaryon emerges, and homokaryons that cannot mate often exhibit persistent rejection.

It is therefore difficult to avoid concluding that mating in basidiomycetes involves overriding the rejection response. Given the probable role of oxidative stress in rejection responses (see above), one important aspect of override may be limitation of oxygen supply to the interior of heterokaryotic cells. This could be achieved by protecting heterokaryotic hyphal boundaries with compounds capable of absorbing oxygen, so making the boundaries better insulated. This fits with the observation that heterokaryotic fungal hyphae do tend to have more distributive and conservative properties than homokaryotic hyphae. It also raises the important general possibility that in terrestrial organisms, sex without protection from oxygen overload is unsafe sex, liable to end in degeneration. Many aspects of the development of terrestrial organisms would then follow. Further grounds for this view, and the way that it may help to explain species formation, will be described at the end of this chapter.

Two further unusual features of basidiomycete mating are that innumerable fusions occur between the participant mycelia and the nuclei from different progenitors do not fuse physically until a late stage, immediately prior to meiosis. Mating between two basidiomycete mycelia therefore implies mating between genomic *populations*, with scope for complex and varied interrelationships to develop. This makes it possible to study a unique kind of genomic or intracellular population biology.

All this contrasts sharply with the typical situation in many other organisms where just two nuclei are combined within a single cell. These nuclei are often derived from recognizably male and female sources and they fuse immediately to produce a unique, diploid "zygote" which then proliferates mitotically (see Chapter

4). Each sexual union in such organisms therefore amounts to a single, unrepeated experiment.

Basidiomycete mating also contrasts with another great group of mycelium-forming fungi, the "ascomycetes". Here, fusion between sexually undifferentiated homokaryotic hyphae typically leads either directly to rejection or to the formation of an unstable heterokaryon which is unable to develop independently for long, before breaking down into its components. However, mating can occur even between strongly rejecting mycelia because sexual union involves specialized sex cells, of which the female is known as an "ascogonium" and the male may be a narrow hypha—an "antheridium"—or a spore. After mating, heterokaryotic "ascogenous hyphae" grow out from the ascogonium and eventually give rise, following karyogamy and meiosis, to a sac or "ascus" containing "ascospores". There is no independently growing heterokaryotic (or diploid) phase of the kind found in basidiomycetes.

Ascomycetes therefore get round somatic rejection by having "private parts" that enable them to keep their sexual and bodily lives separate. In this respect they resemble many other organisms, of which an interesting case is provided by ciliate protists, such as *Paramecium*, which have large macronuclei governing somatic functions and smaller micronuclei which participate in sexual exchange.

There is also a very interesting difference between the genetic systems that control sexual unions in ascomycetes and basidiomycetes. In both groups of fungi there are some species or populations which are non-outcrossing, i.e. capable of "self-fertilization" and hence propagating clonally (cf. Chapter 4). In other cases, sexual outcrossing occurs between organisms that have different mating factors. In ascomycetes there are generally just two such factors or "idiomorphs", each of which must be present to allow mating. Consequently, as in human populations, there are just two kinds of individuals able to mate with one another productively. However, unlike human beings these individuals are often not distinguishable as male and female, because each produces both kinds of sex organs. They are therefore referred to as "mating-types".

Often, individuals of complementary mating-type are described as "compatible", a fact which can cause confusion because such individuals are commonly somatically incompatible due to their genetic difference. Furthermore, somatically compatible individuals having the same mating-type are described as "incompatible". To make matters even worse, individuals that are genetically very disparate may be both sexually and somatically incompatible. Not for the first time, the "paradox" of sex plays havoc with the terms used to describe relationships and it becomes necessary to realize that genetic difference can have very contrasting consequences depending on context. It is therefore very important, even though it may sometimes seem cumbersome, to try to make the context clear when describing the integration or non-integration, the "compatibility" or "incompatibility" of sexual and/or somatic partners.

Nowhere is the difference between sexual and somatic incompatibility more difficult to define than in basidiomycetes. As has been outlined, sexual exchange in these organisms is not mediated through private parts, and so interactions between homokaryons, or between heterokaryons and homokaryons can aptly be described as "somatosexual".

It may therefore be very significant that the genes determining whether sexual acceptance occurs in basidiomycete interactions have numerous alleles. In fact, so great is the polymorphism of these genes that in some species there may be literally thousands of different kinds of individuals able to mate with one another.

The resulting impression that basidiomycetes are the most promiscuous of organisms is reinforced by the fact that any one homokaryon can mate simultaneously with an indefinite number of partners with which it anastomoses at its boundary. Such are the inexhaustible possibilities of an indeterminate body form!

When a homokaryon does mate in this way with more than one partner at a time, it becomes a competition ground for invading nuclei, with the end result that it becomes subdivided into distinct heterokaryotic sectors or "genetic territories". The boundaries between these territories are demarcated by rejection zones due to somatic incompatibility.

A parallel situation occurs in the socially indeterminate swarms of army ants. In all army ants the queens are flightless and are guarded by an entourage of workers; males fly in from other colonies and run the gauntlet of these workers before they can gain access to the queens. The newly mated queens then compete with one another and their maternal queen for possession of the workforce. Only two such queens are ever successful and the parental colony splits into two new and entirely separate colonies.

The association of multiple alleles with sexual acceptance in the somatosexual interactions of basidiomycetes is interesting because it parallels the highly polymorphic nature of the genes responsible for rejection (see above). Also, there are other organisms in which mating involves somatosexual interactions and the genes conferring sexual compatibility are multiallelic. One of these occurs in the slime moulds, where union occurs between amoeboid or flagellated cells, to give rise to a diploid plasmodium (see Chapter 2).

Another example of somatosexual interaction is found in flowering plants. Here, a "male" nucleus contained in a "pollen tube" fuses with an egg cell nucleus in a female "embryo sac". However, before this can occur, a pollen grain must land on a female receptive platform, the "stigma", and germinate to form the apically extending pollen tube which runs the gauntlet of female tissues on its way down a special channel, the "style" (Fig. 7.14). Many pollen grains usually arrive on the stigma, carried there by wind, insects, birds or bats, and these grains give rise to many tubes which race one another to reach the egg cells in the female ovary. Each fertilized egg cell gives rise to an embryo within a seed, so the results of different fertilizations are automatically parcelled up into separate survival or dispersal units. If there are fewer egg cells than pollen tubes, some of the males will inevitably lose out, and, by the same token, the female tissues may impose some "mate choice" or sexual selection.

Many flowers are "hermaphrodite", possessing both a female ovary and male "stamens" consisting of pollen-filled "anthers" borne aloft on supports known as "filaments". The possibility therefore exists for egg cells to be fertilized by pollen from the same flower. However, this situation, which is rather confusingly referred to as "self-fertilization" is often prevented by what is equally confusingly referred to as "self-incompatibility". In fact, the pollen, having been derived by meiosis will be genetically different from, even though related to, the egg cells in the same flower. What is really being prevented, from the point of view of the haploid cells, is therefore "inbreeding" (cf. discussion in Chapter 4). The situation therefore parallels the restriction of inbreeding between homokaryons derived from the same fruit body of a basidiomycete because of the presence of identical mating-type alleles.

The genes which prevent self-fertilization in flowers are known as "self-incompatibility" or "S" genes. Two kinds of self-incompatibility are known. In one kind, "gametophytic self-incompatibility", pollen containing an S allele identical to one of the two S alleles in the diploid style tissue is rejected. In the other kind, "sporophytic self-incompatibility", pollen derived from a parent plant with one or both S alleles in common with those in the style is rejected, even if the pollen itself has an S allele different from that in the style (Fig. 7.14). The result of self-incompatibility is that the pollen tube stops extending, typically in the style (gametophytic incompatibility) or in/on the stigma (sporophytic incompatibility), before it can reach the embryo sac, and degenerates.

As with the mating-type genes of basidiomycetes and slime moulds, the fact that there are many S alleles, such that pollen from another plant is usually accepted, has caused them to be thought of as a device to ensure outcrossing. However, it can also be argued that the enhancement of outcrossing is a secondary consequence of the means by which rejection is overridden, rather than a primary adaptive objective. It is therefore important to understand how rejection is invoked in order to know how it can be avoided.

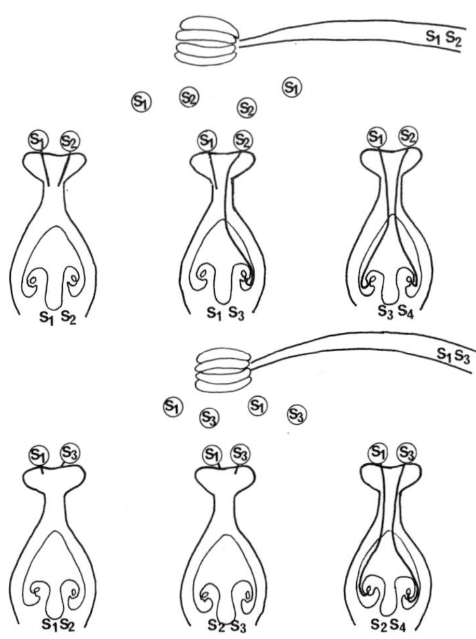

Figure 7.14. Pollen-style-stigma interactions resulting in self-incompatibility in flowers. The top series of drawings of stamen, pollen and ovaries represents gametophytic self-incompatibility, the lower series represents sporophytic self-incompatibility. (Based on Heslop-Harrison, 1978)

Presently it looks as though different kinds of self-rejection mechanisms occur in different groups of plants. In members of the tomato family (Solanaceae) such as tobacco and *Petunia*, which have a gametophytic system, it has been found that the S genes encode "glycoproteins" (carbohydrate-linked proteins) with RNase activity (i.e. they degrade RNA). Expression of the S-RNase within the style tissues is alone sufficient to cause rejection of pollen with a particular S allele. Such pollen is assumed to take in the RNase in some way and so bring about its own destruction; presumably any mechanism which would prevent this would thereby also prevent rejection. In the poppy family (Papaveraceae) a small stylar protein is thought to be recognized by a receptor on the pollen surface and thereby to lead to an increase in intracellular calcium within the pollen tube. By contrast with both these examples, where the rejection reaction is thought to be sited within the pollen, in the cabbage family (Brassicaceae), where the sporophytic system operates, rejection occurs at the boundary of female stigmatic cells.

In the animal kingdom, there are innumerable examples of the delicate "negotiations" required to gain sexual access, and of the sometimes dire consequences of not striking the appropriate chord! Many male human beings must have cringed on first coming across accounts of the dangerous manoeuvres of male black widow spiders or the consuming passion of female praying mantises! More generally, many animals seem to require at least some period of courtship, often involving ritualized displays, chases and alternation between apparently aggressive and submissive behaviour, as a prelude to mating. Males also often appear to vie with one another for access to females, either by showy displays or actual conflict.

Adaptationists have had a field day explaining mate choice in animals as a means of ensuring that only the best quality genes, from the cockiest, healthiest, strongest males gain rights to entry to the next generation (notwithstanding that some of the traits may not be so apt if possessed by females). Courtship has also widely been viewed as the means by which strong "pair-bonds" are formed that will see couples through the rigours and trials of parenthood. However, evolutionarily beneficial though these consequences may prove to be, they also follow automatically from restricted access and the requirement to override rejection.

Apart from understanding how sexual partners do the right thing once they have got together, it is also important to consider how they approach, stay associated and reciprocally or unilaterally invade one another's living space in the first place. Obviously they must attract, bond and open up boundaries to one another in some way. Sexual union involves not just failing to reject but actively accepting a mate.

For organisms with sight, touch, hearing, taste and smell, all these senses may play a role in attracting, detecting, securing and enabling admission to a mate. Perhaps the most universal mechanism, however, and perhaps the one most human beings are least conscious of, is chemical.

Some chemicals act as "signals" or "pheromones" that diffuse out into the environment and cause growth ("chemotropic") or movement ("chemotactic") as well as other responses by potential mates. These chemicals are commonly isoprenoids (sesquiterpenes, steroids and apocarotenoids) or hydrophobic peptides and are produced by many groups of organisms, ranging from fungi to mammals.

Pheromones can induce responses at very low concentration, a fact that makes it possible for us to use them practically to lure male pests (of other organisms, such as tsetse flies, that is!) to their doom! With this in mind, it is intriguing to note that the sesquiterpene compound which attracts male "zoospores" of the fungus *Allomyces* to female sex organs is called "sirenin", after the mythical creatures that lured sailors with their sweet song!

The role of peptides as pheromones has been particularly well-studied in certain fungi that exist for all or part of their lives as yeasts (i.e. in a budding or dividing one celled form). In the baker's yeast, *Saccharomyces cerevisiae*, there are two mating-types, known as a and Â. These each produce specific peptides, a-factor and Â-factor respectively, which induce reciprocal responses in cells of the other mating-type. These responses cause the cells to elongate or "shmoo", thereby forming a conjugation tube, and to arrest at G1 in the cell-division cycle (see Chapter 4). The mating-type genes regulate but are not themselves the genes that encode a and Â-factor or the receptors to which these factors bind. However, in yeasts produced by basidiomycete fungi, peptide pheromones are coded for directly by mating-type genes and there are indications that one of the multiallelic mating-type genes of mycelium-forming basidiomycetes may also do so.

One interesting consequence of the presence of peptide and steroidal systems in fungi is the possibility that these organisms may be affected by the hormonal systems affecting gender-specific attributes in mammals. In fact, there is growing evidence that fungi are affected in this way, possibly explaining the different susceptibility of male and female human beings to fungal diseases.

Whilst pheromones may bring mating partners together, they do not in themselves cause them to stay together. In cell-to-cell interactions, bonding has commonly been found to be due to the presence of glycoproteins which attach to each other as receptor and ligands. In the yeast, *S. cerevisiae*, these take the form of mating-type-specific "agglutinins", which cover the surface of conjugation tubes.

Finally, the fact that some mammals are able to detect genetic differences in the products from MHC genes (see above) raises the fascinating possibility that these genes could play a role in mate choice. There is evidence that male mice are more liable to select female mice with different MHC alleles from themselves as mates. There is also evidence that the smell of a new male partner may be sufficient to cause rejection of an embryo conceived following fertilization by a previous partner. Such findings provide a fascinating parallel with the multiallelic mating systems of basidiomycete fungi and slime moulds, as well as with the "self-incompatibility" systems of plants.

3.2. Infancy

People, whether male or female, often have a curious reaction to anything that they perceive to be "sweet" or "cute", wanting to pick it up (if possible), cuddle, fondle and nurture it. Many pets have made themselves a living out of this tendency!

It is difficult to define what it is that elicits this reaction, though a lack of threat, fuzzy boundaries and an air of utter self-insufficiency may all play some part. The reaction might therefore seem to result from a genetically pre-programmed desire for parenthood that just happens to get displaced (sometimes at the expense of human babies) onto other creatures. However, this explanation may be rather one-sided, focusing more on the receiver than the transmitter of signals.

In general, any dependant being must be accepted by its "carer" if it is to survive. It must not, therefore, cause dysfunction or be recognized in any way as a threat, whether at molecular, cellular or larger scales of organization. It should, on the other hand, be malleable—able to fit in. The great evolutionary success (for the time being) of domesticated animals, based as some would say on neoteny (see Chapter 5), illustrates the point. Perhaps the human response to imagined deities, with whom atonement (at-one-ment) can be gained through self-effacement, can also be interpreted in these terms.

The young of mammals illustrate many of the kinds of mechanisms that enable infant beings to be accepted. It all begins at the time of implantation (if not at conception, when male sperm have to be accepted from males which have to have been accepted by females!)—when the young embryo becomes embedded in the wall of the womb. From this stage on, the developing tissues and vascular infrastructures of the foetus become increasingly integrated with those of the mother, through the interactive interface of the placenta, until at last, after birth, all three dissociate.

There is a profound problem in all this: why shouldn't the foetus, which is genetically different from its parents, be rejected by the mother's immune system? After all, in later life an offspring will reject a transplanted organ from its parent and *vice versa*. Basically, it seems that if the pregnancy is to go full term, the mother's rejection response must be blocked. The mechanism by which this is achieved is still uncertain, but there is some evidence that it may actually depend on genetic difference, just as does the acceptability of another mating-type or pollen tube (see above).

Another interesting fact is that embryonic tissues are immunologically immature, so that cells or tissues from other organisms that would be rejected in later life are accepted. This fact led some time ago to the development of a technique whereby chimaeric animals, half sheep and half goats (i.e. geep and shoats) were produced by mixing embryonic cells from two different sources at an early stage of development.

Such acceptance of foreign cells or tissue encountered early in development is known as "acquired immunological tolerance". It may involve the same kind of "learning process" that prevents a mature animal from rejecting its own tissues. This learning process is believed to involve a combination of positive and negative selection during T cell development in the thymus gland. T cells with receptors able to recognize not yet extant foreign peptides complexed with self MHC molecules (see above) are positively selected, whereas T cells that react strongly with self MHC molecules, or self MHC molecules complexed with extant peptides undergo cell death.

Once born, a young mammal is safe from the potential depradations of its mother's immune system, and may even gain protection against infection from antibodies in milk. At this stage behavioural mechanisms take over that reinforce its mental attachment to and prevent rejection and even attack by its parents, allowing it to be nurtured. Ultimately, as it becomes more able to lead its own life, the bonds to its parent(s) (the proverbial "apron strings") start to fray as the tension between its own demands and parental rejection ("parent-offspring conflict") increases. When the bonds finally sever, a young adult emerges.

3.3. Kin Selection

With the discounting of group selection as a major evolutionary mechanism (see Chapter 1), some alternative genetic explanation for apparently self-sacrificing behaviour in populations became needed.

An obvious example is when a parent does not itself directly benefit, and may indeed suffer loss, through caring for its young. Here there may yet be an overall increase in the survivorship of parental genes (known as "inclusive fitness") if the disadvantage to the parent is more than counteracted by the advantage to its offspring. There may, however, come a point where this overall benefit begins to diminish as offspring become more independent. This provides an explanation for parent-offspring conflict (see above) in terms of evolutionary genetics.

More generally, any gene which is disadvantageous to an individual may increase in frequency if it confers a sufficient advantage to the bearer's relatives, i.e. is subject to "kin selection". For example, in family groups of determinate diploid organisms, a trait which halves the chances of genetic survival of a parent but more than doubles the chances of survival of a daughter or quadruples the chances of a niece will, theoretically, increase in frequency.

According to kin selection theory, altruistic behaviour might therefore commonly be found in kin groups maintained by inbreeding. However, inbreeding itself can have genetic costs, especially in diploid organisms, and so the maintenance of kin selection may present difficulties. Also, like many other theories of evolutionary fitness based on genetic determinism, kin selection theory ignores the effects of boundary constraints or context. It may therefore paint a partial picture based on the perspective gained from only a single standpoint, that of the gene as a fully particulate unit of selection. Even so, the ability of rodents to sniff out one anothers' MHC specificities implies that the possession of common MHC alleles is important to the establishment of some forms of co-operative behaviour.

3.4. Infiltration

In discussing sexual union and infancy, I focused on the processes by which nearself entities, members of what are usually considered to be the same species, evade rejection and gain acceptance.

An essentially similar set of processes allows entities that are genetically highly disparate to become associated for greater or lesser periods of time in diverse kinds of symbioses. As discussed in Chapter 1, these symbioses may be neutralistic—with neither associate benefitting directly at the expense of or to the advantage of the other, parasitic—with one benefitting at the expense of the other, or mutualistic—with both benefitting.

In parasitic symbioses and also neutralistic symbioses that are indirectly costly for a "host", there is an evolutionary pressure for invaders to be detected and rejected. The ability of invaders to gain access across and proliferate within the boundaries of a host without being rejected then depends on infiltration.

Infiltration can be thought of as a kind of "Trojan Horse" or "guerrilla" strategy that avoids immediate confrontation with the host. It has two, interrelated components. The first component, which is purely physical, involves being organized on a scale which is smaller or more diffuse than the host system. This allows gaps and channels in the host system's boundaries to be located and penetrated more readily. The second is "stealth". This can involve doing nothing by way of emitting vocal, visual, chemical or tactile "signals" that might allow detection. Alternatively, or additionally, positive measures may be adopted that disguise boundaries, prevent transmission or reception of signals, or deliver confusing information.

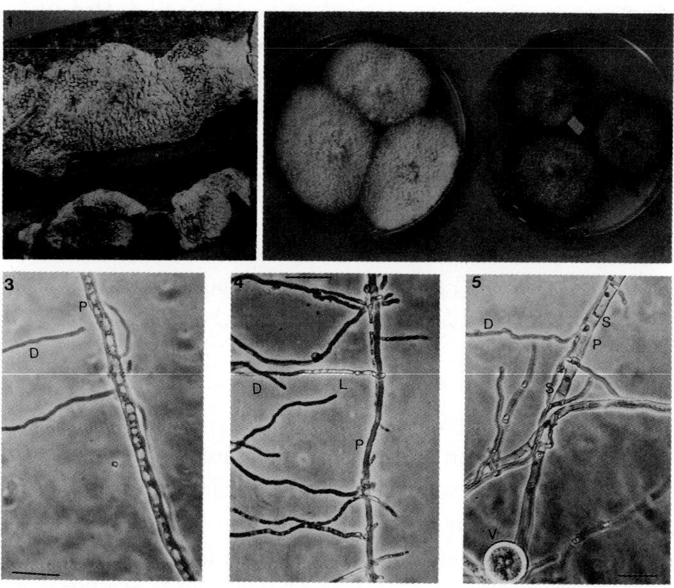

Figure 7.15. "Guerrilla warfare": infiltration of one basidiomycete fungus species, *Datronia mollis*, by another species, *Phanerochaete magnoliae*. (1) Two lengths cut from a beech branch, the lower of which bears normal fruit bodies of *D. mollis*, whilst the upper length bears fruit bodies through the tubes of which *P. magnoliae* has formed spore-producing spines. (2) Mycelia of *D. mollis* growing in Petri dishes. The dish on the right has been inoculated centrally with *P. magnoliae*, whilst the other dish has not. Notice the inhibition and pigmentation of the *D. mollis* colonies in contact with *P. magnoliae*. (3-5) Responses of hyphae of *P. magnoliae* (P) and *D. mollis* (D) involving entwining, lysis (L, often associated with increased refractility), septal partitioning of evacuated compartments (S) and formation of protoplasm-filled vesicles (V) from which *P. magnoliae* regenerates. Scale bars represent 25 μm. (From Ainsworth and Rayner, 1991)

A striking example of infiltration is provided by a wood-inhabiting basidiomycete fungus, *Phanerochaete magnoliae* (Fig. 7.15). This fungus has recently been collected from several separate locations in England. On each occasion its fruit body has taken the form of a collection of spore-producing spines protruding from the spore-producing tubes of the fruit body of another wood-inhabiting basidiomycete, *Datronia mollis*. When the mycelium of *P. magnoliae* was grown in culture, it was found to produce a rapidly extending, very sparsely branched system which readily penetrated across the much more densely branched mycelium of *D.*

mollis. Wherever *D. mollis* hyphae made contact with *P. magnoliae* hyphae, mutually destructive hyphal interference reactions were provoked until eventually *P. magnoliae* spore-producing surfaces emerged, phoenix-like, from localized centres amidst the chaos. A better demonstration of the principles of guerrilla warfare would be hard to find!

Two other wood-inhabiting basidiomycetes, *Pseudotrametes gibbosa* and *Lenzites betulina* exhibit temporary parasitism when gaining access to wood in which species of *Bjerkandera* and *Coriolus* have established. Their hyphae are able to invade mycelia of the latter fungi without any sign of causing a reaction. They then encoil and parasitize the host hyphae, eventually replacing them altogether and taking over their territory.

This behaviour resembles that of temporarily parasitic ants and strangler figs. Queens of temporarily parasitic ants invade colonies of a host species, kill the resident queen and take over her workforce. The parasitic queen's offspring are then cared for by the host workforce until eventually they come to predominate. Strangler figs climb and encoil trees as a means of gaining access to the forest canopy where they spread out their own assimilative greenery. Their stems then thicken around the host which eventually dies and rots.

Many fungi are "biotrophic", drawing resources from living cells or tissues of plants, animals or other fungi. As mentioned earlier in this chapter, fungi that are biotrophic plant parasites depend on not setting off a hypersensitive reaction in their hosts. If they succeed in this respect they are able to tap into the host system, acting as sinks that draw off resources in much the same way as would, for example, an enlarging fruit. In the case of biotrophic fungi which colonize leaves, for example, a common effect is the formation of green islands in the infected regions whilst adjacent tissues senesce prematurely and yellow as their resources are mobilized towards the fungus (Fig. 7.16). Other biotrophic fungi commonly cause various distortions of growth or division in infected tissues, causing such symptoms as swollen stems, leaf curls and witches' brooms. Such effects probably involve changes in plant hormone concentrations produced or induced by the fungus.

One of the hallmarks of biotrophic fungi is the production of highly specialized hyphal branches or branch systems, known as "haustoria" (Fig. 7.17). These form an intimate interface with plant cells. They breach the plant cell wall by means of a narrow penetration tube and then enlarge or proliferate within this boundary without penetrating the plant cell membrane—rather like using a glove box. The fungus then draws resources from the plant cell and, *via* connective plasmodesmata, from its neighbours. The fungus is therefore well and truly tapped in, becoming, in effect, a part of the plant's infrastructure but proliferating its own rather than the plant's genes.

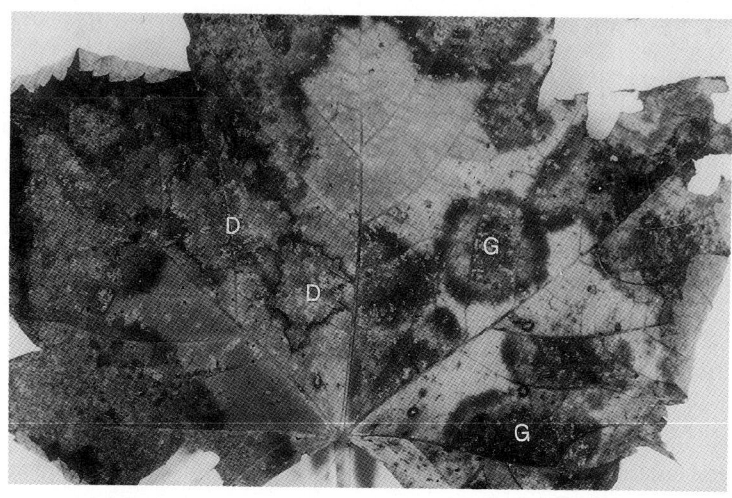

Figure 7.16. "Green islands" (G) and dead patches (D) in a sycamore leaf resulting respectively from "biotrophic" and "necrotrophic" infection by the fungi *Uncinula bicornis* and *Cristulariella depraedens*.

The ability of a biotrophic fungus to become established in this way has been found to be determined by genetic systems, both of the fungus and the plant. When the fungus sets off a hypersensitive reaction, its interaction with the plant is described as "incompatible", and the fungus itself is said to be "avirulent" (incapable of causing disease). When the fungus establishes an infection, the interaction is described as "compatible" and the fungus as "virulent".

In some cases a "gene-for-gene" system has been discovered, such that incompatibility ("resistance") is only expressed if a specific gene for avirulence in the fungus interacts with a specific gene for resistance in the host. Perhaps not surprisingly, in view of what has already been said in this chapter, the resistance genes are highly polymorphic, occurring both at several different loci and in multiallelic form. Fungal avirulence genes on the other hand appear to occur at many loci, but there is less evidence for multiallelism.

The precise mechanisms responsible for compatibility and incompatibility are still not certain. However, a plausible model is that a fungal avirulence gene product is "recognized" in some way by a resistance gene or gene product. This event triggers a cascade of processes culminating in the production of phenolic and isoprenoid compounds and cell death. In compatible interactions this process is prevented either by the absence of the fungal gene product or by suppression of recognition or subsequent events.

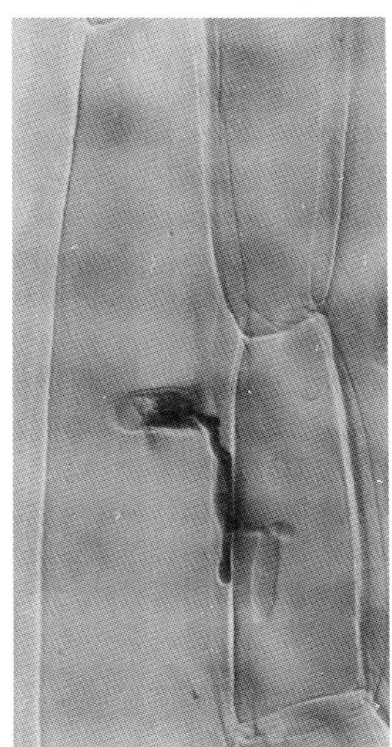

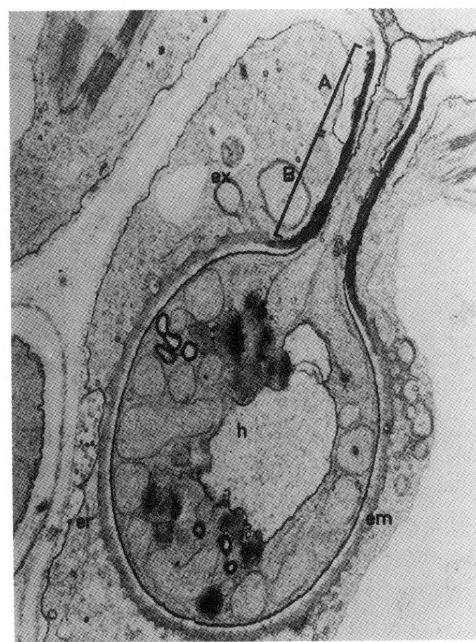

Figure 7.17. Fungal haustoria. On the left is a light microscope photograph showing a spore (partly dark contents), germ tube (dark contents) of a powdery mildew fungus on the surface of a barley leaf. An infection peg has penetrated from the germ tube through one of the plant cell walls and then enlarged to produce the haustorium (light contents). (Courtesy of S.A. Archer). On the right is an electron micrograph through a haustorium (h) of the fungus *Albugo candida* in a host plant cell. The cell envelopes of the fungus and host remain distinct except where they are fused in a neck-ring (A,B). (From Woods and Gay, 1983)

A kind of "social biotrophic parasitism" is exhibited by "inquiline" species of ants, the queens of which lack their own workers and so are completely dependent on the workers of a host species to support them and their offspring. There appear to be a variety of ways in which the inquiline queens infiltrate and become accepted into host colonies. For example, *Anergates atratulus* appears to play dead when met by a worker of the host species, *Tetramorium caespitum*. She then clings onto the worker's antenna with her jaws and is dragged into the nest. *Epimyrma stumperi* either plays dead or captures a *Leptothorax tuberum* host worker, rubs it with her forelegs and grooms herself, thereby perhaps acquiring the host's colony odour.

Teleutomyrmex schneiderii, on the other hand, seems to appease workers of *Tetramorium caespitum* with glandular secretions. Perhaps the most remarkable case is provided by *Leptothorax kutteri*, which has recently been shown to evade attack by workers of its host species, *Leptothorax acervorum*, by producing a secretion from her Dufour's gland known as a "propaganda substance". This substance causes host workers to attack each other, and so overrides nestmate recognition in the host colonies.

A kind of "molecular-scale biotrophic parasitism" occurs amongst certain algae and fungi. The parasitic red alga, *Janczewskia gardneri*, transfers its nuclei through specialized "conjunctor cells" into the cytoplasm of cells of its taxonomically closely related host, the seaweed *Laurencia spectabilis*. Following transfer, the parasitic nuclei undergo division whilst the division of host nuclei is arrested, culminating in a situation where the former may outnumber the latter by as much as twenty-to-one. Similar kinds of processes occur during parasitism by the mucoraceous fungi, *Chaetocladium* and *Parasitella* on their host fungi.

Certain fungi, known as "endophytes", reside within the tissues of living plants without causing any visible symptoms. Until relatively recently, they have been regarded as innocuous and insignificant. However, it has now been recognized that endophytic fungi inhabit virtually all larger plants and that they can confer benefits, such as protection from herbivores and parasites, as well as potentially becoming very destructive if the plant becomes subjected to stress. Whilst they develop sparsely, and often virtually undetectably, in healthy plant tissues, they switch into an actively exploitative mode if the tissues become dysfunctional. Some examples of wood-inhabiting versions of these organisms were mentioned in Chapter 6, where they were described as "specialized opportunists".

There is a very fine borderline between establishing as an endophyte and capitalizing on dysfunction brought about by some other agency, and actually causing dysfunction and so being pathogenic. This is emphasized not only by the activation of exploitative phases in stressed plants, but also by the occurrence of fungi that in infiltrating the interior of plants actually trigger a "suicidal" reaction by their hosts. The latter situation is illustrated by fungi causing diseases known as vascular wilts.

Vascular wilt fungi gain access to the water-conducting tissues of xylem where they typically proliferate as yeasts or yeast-like forms that are readily distributed by the water flow to the canopy. Their presence in the xylem induces boundary-sealing processes, such as production of gums, phenolic compounds and tyloses, that block the conduits and so deprive the canopy of water. Infected plants therefore wilt and die. A notorious example is Dutch elm disease.

Infiltrative development is also found in fungi that cause diseases in animals. Many of these fungi can switch between yeast-like forms which are readily distributed through vascular systems and mycelial forms which ramify through tissues.

Many of the fungi which cause disease in human beings only become damaging, sometimes after remaining latent for years, as a result of some form of debilitation, such as may be caused by immune deficiency, drug-taking, trauma and old age. Some of these fungi possess mechanisms which enable them to evade the immune system. For example, *Cryptococcus neoformans*, which is harboured in pigeon droppings, produces yeast cells that are covered in a polysaccharide capsule. Other fungi, such as *Candida albicans*, can actually cause immunosuppression. In fact, the need to evade or suppress the immune system is a general requirement for inhabitants of vertebrate tissues. For example, a mechanism found in trypanosomes, the microorganisms that cause sleeping sickness, actually involves rapid changes in their surface coating of antigens as a result of somatic recombination mechanisms not unlike those underlying antibody diversity itself.

3.5. Reciprocity

Non-mutualistic symbioses are potentially unstable both because they can lead to the elimination of one partner (which is bad news for a dependent parasite as well as its host) and because they lead to arms races (see Chapter 1). The gene-for-gene relationships between biotrophic plant parasites and their hosts is probably due to such arms races.

For partnerships to be stable for long periods, the development of interdependence or reciprocity is necessary, such that each participant fulfils a need in the other. In the case of green plants and fungi, reciprocity is due to the fact that whilst plants, being photosynthetic, do not need to absorb organic sources of carbon, they do need to absorb mineral nutrients and water. Fungi, on the other hand need supplies of organic carbon and, in having a mycelial organization are highly effective in locating and absorbing mineral solutions. Providing that it is appropriately placed, for example growing below ground rather than on aerial surfaces, a biotrophic fungus can therefore be an absorbtive accessory to a plant.

This is precisely the situation which occurs with many mycorrhizal fungi; the fungus explores soil far more effectively than can root hairs and so greatly enhances the ability of infected root systems to take up water and mineral nutrients. On the other hand, the fungus can draw organic resources from the plant as a result of the intimate connections that it forms with root cells and tissues.

Three kinds of mycorrhizal (fungus-root) associations generate the infrastructures responsible for supporting the major vegetational systems of the world. "Sheathing" or "ectomycorrhizas" (Fig. 7.18), are most characteristic of temperate and boreal forest, where they are produced by members of the Pinaceae (e.g. pines, spruces, larches and firs), Betulaceae (birches), Salicaceae (willows), Fagaceae (e.g. beeches and oaks) and Myrtaceae (eucalypts). They can also occur in arctic-alpine communities, e.g. in association with "mountain avens" (*Dryas octopetala*) and in tropical forests, e.g. with the Caesalpinoidae. "Vesicular-arbuscular" or "VA" mycorrhizas are formed by a great many plants, and indeed there is fossil evidence that they accompanied, and were perhaps pre-requisite for the first invasion of land by vascular plants. They are abundant in tropical forest and grassland communities of all kinds, and may be particularly important in agricultural crop plants. "Ericaceous" mycorrhizas are formed, as their name implies, by members of the heather family (Ericaceae), and are fundamental to the development and functioning of moorland and heathland communities.

In ectomycorrhizas, the fungal partner forms a compact layer of densely proliferated hyphae around the outside of absorptive "short roots" (see Chapters 4 and 5). From the inside of this layer or sheath, fungal hyphae ramify amongst the cells of the root "cortex" forming a network known as the "Hartig net". This network is the "communications interface" between fungus and plant, and within it fungal and plant cell wall boundaries become merged in an "involving layer". From the outside of the sheath emanate diffuse or aggregated "foraging" systems (see Chapter 6) of fungal hyphae. These systems are generally produced more abundantly in poor than in rich soils and allow the fungus both to distribute itself from root to root and to gather soil solution which can be passed on to the plant partner.

A great many fungi can form ectomycorrhizas. Many of them, for example such genera as *Amanita*, *Boletus*, *Cortinarius*, *Hebeloma*, *Russula*, *Lactarius* and *Suillus* are "agarics", which produce "mushrooms" and "toadstools" as fruit bodies. Others include the true and false truffles, earthballs and false morels. Douglas fir (*Pseudotsuga menziesii*) is thought to be capable of forming ectomycorrhizas with as many as 2,000 different species of fungi.

The degree of specificity between plant and ectomycorrhizal fungi is of great interest and consequence. Since the foraging components of mycorrhizal fungi form connections between neighbouring roots, it follows that these fungi can serve to interconnect adjacent plants, forming a living bridge through which resources are exchanged from one to the other. The resultant fungus-mediated communication between neighbouring plants may be a very important component of plant community functioning (see Chapter 2). The extent to which such communication occurs will be determined by the readiness with which the same fungus can form mycorrhizas with separate plants of the same and different species. In parasitic relationships between fungi and plants, the plant may be expected to benefit by being able to recognize and reject the fungus, so giving rise to highly specific, even gene-for-gene, relationships. On the other hand, to reject a potential ally like a mycorrhizal fungus might seem unadaptive, and so at least in theory mutualistic relationships have been considered likely to be fairly catholic. Should this prove to be true, then the potential range of influence of the infrastructures formed by mycorrhizal fungi in plant communities could be very great.

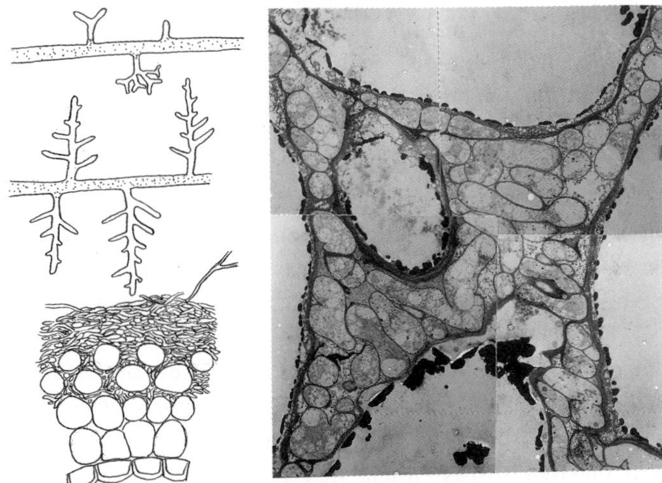

Figure 7.18. Structure of ectomycorrhizas. Left, top and centre: long (stippled) roots of pine and beech respectively bearing forked and laterally branching mycorrhizal short roots. Left, bottom: cross-section through part of an infected short root showing external sheath of compactly arrayed hyphae and "Hartig net" spreading between cortical cells of the host. Right photograph: section through part of an ectomycorrhiza formed by the fungus *Rhizopogon roseolus* with larch, viewed under the electron microscope. The intercellular spaces between relatively large and empty cortical cells of the host are packed with fungal tissue forming the Hartig net. (Courtesy of K. Turnau)

Unfortunately, although some studies have shown that a particular ectomycorrhizal fungus can indeed associate with plants of different species, there is generally too little information about how specificity within and between fungal and plant species and populations varies. On the other hand, and to the surprise of some people, evidence has been found that certain plants can be incompatible with, i.e. reject particular fungal partners, associated with the production of phenolic barriers.

The occurrence of rejection between plants and ectomycorrhizal fungi highlights the potentially complex issues underlying the formation and outcome of mutualistic partnerships. As in human attachments, just accepting the first prospective partner that comes along need not mean living stably together forever after.

Depending partly on genetic make-up and partly on circumstances, some partners may give or take more than others, and mycorrhizally interconnected plants can compete as well as co-operate with one another through fungal networks. Any particular relationship may change from mutual to one-sided if conditions change. Ectomycorrhizal fungi are a considerable drain on resources, using up as much as a quarter of photosynthetic production. They therefore only yield a positive return on investment under conditions where the plant partner is self-insufficient—as in soils which are moderately poor in mineral content. Addition of fertilizers to soil has, correspondingly, been shown to reduce the formation of mycorrhizas.

Neither should the outcome of any particular relationship be judged purely in terms of the exchange of resources between partners. Over the years it has been found that mycorrhizal fungi may affect the microenvironment around host roots in many ways. For example, they may alter aeration conditions; they may impede or enhance mineral nutrient release from organic matter (mineralization); they may impede disease-causing organisms; they may sequester toxic metals that would otherwise be taken up from soil into the roots.... all to varied extents in different fungi.

In VA mycorrhizas, the fungal partner is a member of the Zygomycetes, the pin-moulds and their allies. The hyphae grow loosely over the outside of infected roots, rather than forming a compact sheath. Branches from the superficial hyphae penetrate into the root tissues, where they both spread between and invade the interior (hence sometimes being referred to as "endotrophic") of the plant cells. Within the plant cells, highly branched haustoria are formed which look like miniature bushes ("arbuscules") when viewed microscopically. These represent the nutritional interfaces between plant and fungus, and like the haustoria of biotrophic parasites do not, indeed must not, penetrate the plant cell membrane. In addition to the arbuscules, flask-shaped "vesicles", thought to function as storage organs, are produced by the fungus both within and between the plant cells.

As with ectomycorrhizal fungi, VA hyphae growing in the soil outside infected roots gain access to soil solution and can form connections with neighbouring plants. These external hyphae are thought to play an important role in binding soil particles; their role in sand-dune formation was mentioned in Chapter 6. Specificity in VA mycorrhizal associations is thought to be very low, so that formation of widespread infrastructures in plant communities is possible. However, the connective mycelium of VA fungi may not have the spatial range possessed by the cable-like mycelial cords formed by some ectomycorrhizal fungi.

Ericaceous mycorrhizas differ from ecto- and VA mycorrhizas in that they produce little external mycelium. Rather, the fungus (which is commonly a particular ascomycete species, *Pezicula ericae*) occurs almost entirely within the cells of the root cortex (Fig. 7. 19), where it may account for as much as 70 per cent of the weight of the root. This may be related to the fact that their host plants grow in persistently or temporarily waterlogged soils and have very finely and densely branched roots ("hair roots") bathed in colloidal soil solution.

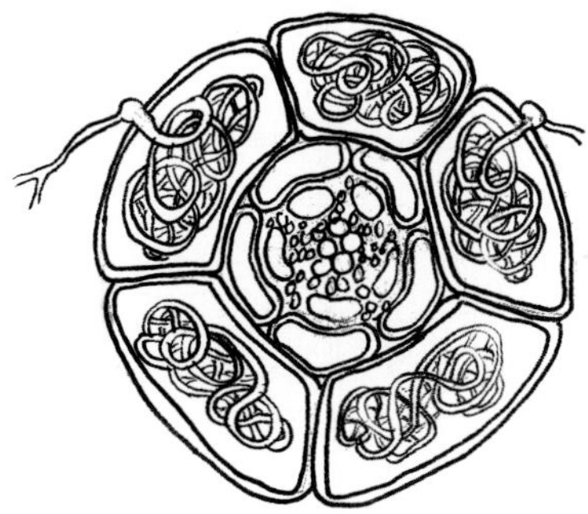

Figure 7.19. Structure of ericaceous mycorrhizas, as seen in a cross-section through a "hair root" of ling heather. The root has a single layer of large cortical cells, each filled with encoiled fungal hyphae, surrounding the vascular cylinder.

Fungi can also form reciprocal relationships with animals. Some of the best known examples involve insects that actually cultivate fungi in so-called "fungus gardens" (see Chapter 2). However, the example which perhaps most strikingly illustrates the complexities and delicate balances within reciprocal relationships is provided by members of an extraordinary family of basidiomycetes, the Septobasidiaceae.

The Septobasidiaceae contains two genera, *Septobasidium* and *Uredinella*, that form close relationships with scale insects. In the most elaborate cases, e.g. in *Septobasidium burtii*, the fungal mycelium houses a colony of the insects on the surface of tree bark, protecting them from the vicissitudes of the external environment, including the attention of parasitic wasps. Some members of the colony are able to move about, but others are immobilized as living bridges between the fungus and the tree (Fig. 7.20). The immobile individuals have penetrative tubes, "stylets", which tap into the tree's medullary rays, but these individuals are themselves infiltrated by the fungus which produces coiled assimilative hyphae within their tissues. The insects have a degree of freedom, but only at some sacrifice to the "fungus god" that shelters them whilst gaining access to the tree.

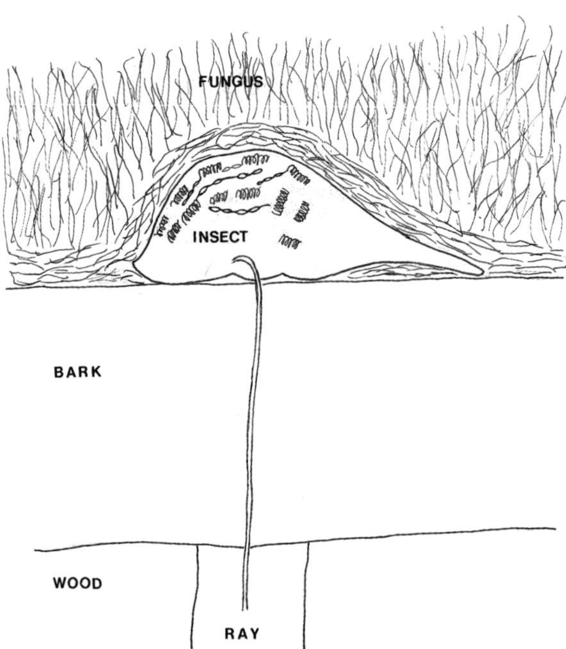

Figure 7.20. Diagram illustrating the sacrificial partnership formed between the scale insect *Aspidiotus osborni* and the basidiomycete fungus, *Septobasidium burtii* infesting tree bark. The fungus produces hyphae consisting of assimilative coils and chains of interlinked spindle-shaped compartments within the blood cavity of the insect, which in turn inserts a feeding tube ("stylet") through the bark into a medullary ray.

4. Overcoming Rejection—Brute Force

Unlike infiltrative mechanisms, which depend on some means of avoiding immediate confrontation and hence provide some opportunity to establish co-operative relationships, overcoming rejection involves the wholesale destruction of boundaries.

Overcoming rejection therefore entails applying considerable force, and dissipating large amounts of energy, at the interface between opposed systems. With the exception of enslavement, it also yields no gains other than resources and/or territory. Nevertheless, mechanisms that overcome rejection are widespread in living assemblages of all kinds and scales. These mechanisms can be classified into three types, depending on whether the assailant is organized on the same, or on smaller or larger scales than the assailed.

4.1. Obliterating Neighbours

Adjacent entities that do not integrate but share resource requirements and are organized on the same scale within some arena are, by definition, both competitive and incompatible. Such entities may co-exist, after a fashion, if the self-protective and defensive mechanisms deployed at their boundaries are maintained in balance, so leading to prolonged deadlock—as in the population and community mosaics of fungi and lichens (see Figs 2.21, 6.17). However, any imbalance between the abilities of the entities to reject and obstruct one another is liable to result in considerable instability. Providing that it is still able to assimilate resources, the better shielded and/or better armed entity can turn defence to offence and so gain access to the domain of its "weaker" neighbour, which is dispossessed and displaced.

Examples of such territorial displacement at molecular and cellular organizational scales have already been mentioned; however, conflict between neighbours is more usually thought about in terms of interactions between organisms. Here some of the most striking illustrations of the principles and processes involved, particularly in indeterminate systems, are found in basidiomycete fungi. As mentioned in Chapter 6, these organisms often predominate at closed stages of fungal community development and correspondingly possess relatively combative ecological strategies.

Some examples of displacement outcomes from encounters between different species of basidiomycetes paired against one another in laboratory culture are shown in Fig. 7.21. Similar patterns of displacement can occur between mycelia of the same fungal species, although mutual exclusion is the more usual outcome of such interactions, unless the combatants are very disparate.

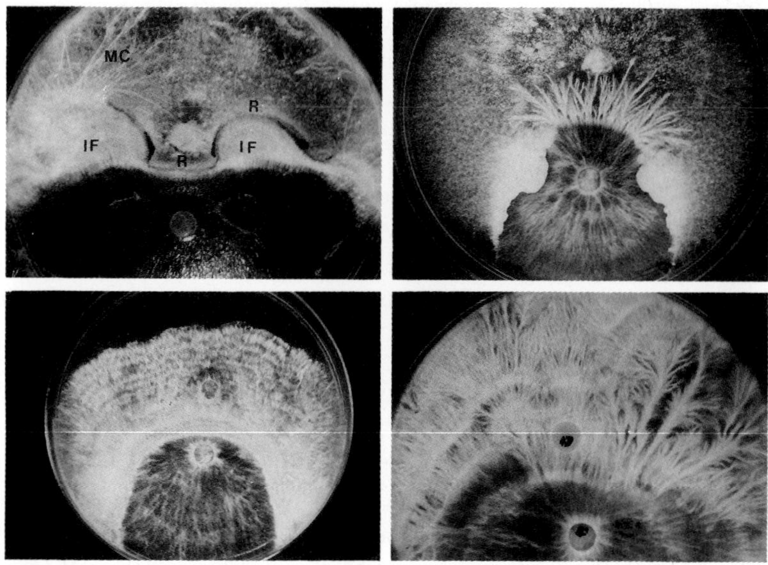

Figure 7.21. Brute force in interactions between mycelia of different species of basidiomycete fungi when grown next to one another in Petri dishes. (Top left) *Phanerochaete velutina* (lowermost) against *Phlebia radiata*. Periodically extending invasion fronts (IF) produced by *P. velutina* have halted where *P. radiata* has formed ridges (R) of aerial mycelium, but these ridges have been breached by penetrative mycelial cords (MC). (Top right) *Hypholoma fasciculare* (lowermost) against *Coriolus versicolor*. *H. fasciculare* has produced invasive mycelial cords at the "weak" centre of the interaction interface, but in so doing left itself open to a flanking "pincer attack" by dense invasion fronts produced by *C. versicolor*. (Bottom left) *Hypholoma fasciculare* (lowermost) against *Peniophora lycii*. *H. fasciculare* has produced a sheet-like invasion front consisting of a series of bands of emergent mycelium. Close examination of the invasion front reveals that major bands of equal width can be subdivided into 2, 4, 8 and possibly 16 minor bands of dense mycelium alternating with zones of sparser mycelium. (Bottom right) *Hypholoma fasciculare* (lowermost) against *Phlebia rufa*. *H. fasciculare* has invaded *P. rufa* both in the form of mycelial cords with delta-like branching (right) and as more diffuse bands showing complex variations in hyphal density. (From Griffith et al., 1994a,b).

Viewing these encounters it is difficult not to be reminded of engagements between territorial armies as the mycelia thrust, counterthrust and redistribute resources to and from their interface. However, unlike conflicts between human armies, there are no generals supervising events. Instead the dynamic and complex patterns arise from feedback mechanisms that result in the opening, sealing and repositioning of the boundaries of the indeterminate, collectively organized and hence fluid-like mycelial systems.

For all that it might seem to involve intelligent scheming, the interplay unfolds automatically, without pre-calculation or masterminding, as the outcome of counteraction between the four fundamental processes described in Chapter 6. Like the result of a football match, this outcome may alter fundamentally as a consequence of small changes in initial conditions whose effect is impossible to predict with foresight but easier to rationalize with hindsight. For example, in one study, when the same interaction between *Coriolus versicolor* and *Peniophora lycii* was repeated under as identical conditions as was possible within the range of experimental error, three strikingly different results were recorded (Fig. 7.22).

A deciding factor in these interactions is the relative ability of the combatants to produce stationary "aerial barrages" or invasive mycelial cords, sheets and fans. Being well insulated, these emergent states are relatively resistant to potentially damaging physical or chemical agents, whilst being able to overwhelm the opposing system.

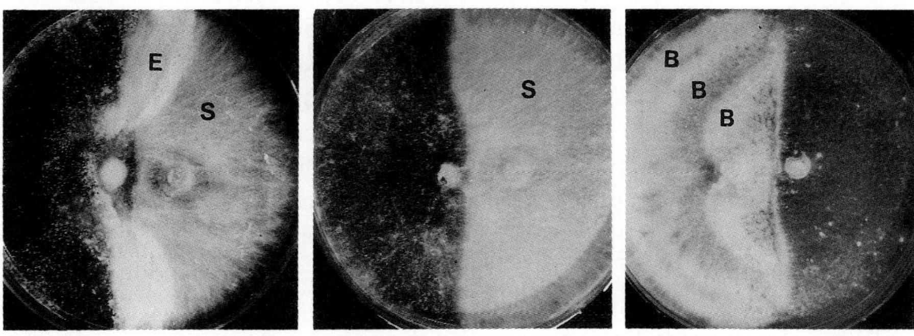

Figure 7.22. Different outcomes of the interaction between mycelia of the basidiomycete fungi, *Coriolus versicolor* and *Peniophora lycii*, when cultured next to one another in a Petri dish. *Coriolus versicolor* (inoculated left) has invaded and replaced *P. lycii* with diffusely spreading mycelium (S) in the left and centre dishes, either directly or preceded by the emergence (E) of a dense mycelial band at the interaction interface. However, *P. lycii* has replaced *C. versicolor* by means of a series of bands (B) of invasive mycelium in the dish shown on the right. (Courtesy of G.S. Griffith)

Extraordinarily parallel territorial patterns are produced by other indeterminate life forms in which penetration and overtopping provide effective means of displacement. Stony corals, for example, produce "sweeper tentacles" and "sweeper polyps", loaded with nematocysts (stinging cells), with which they inflict damage on opposing colonies. The "hyperplastic stolons" of the hydroid *Hydractinia echinata*, especially, resemble the emergent hyphae of basidiomycetes, even though being multicellular rather than cellular structures (see Chapter 2). These stolons emerge clear of the substratum at the interaction interface and produce and discharge successive layers of nematocysts, so clearing a way into the opposing colony.

In the case of systems consisting of assemblages of determinate entities, the relationship between the size and number of individuals arrayed at battlefronts may have an important bearing on the outcome of conflicts.

This relationship has recently been modelled in ants by means of a series of coupled differential equations used early in the twentieth century by an engineer called Lanchester to describe the rates of attrition of two opposing human armies. Lanchester examined a variety of combat scenarios in which he examined the relative importance of the number and fighting value of individual units. In one scenario the units take one another on in a series of one-to-one duels, so that superfluous units on the majority side remain unengaged until enemy individuals are available for single combat. Here, Lanchester's linear law applies, with the overall fighting strength of each army simply proportional to the product of the number and fighting value of individual units. In another scenario, all combatants on both sides were considered to be equally vulnerable to attack from every combatant on the opposing side. Each combatant on the numerically weaker side would thereby be subject to attack by more than one combatant from the opposing side, resulting in a disparity which is prone to amplify with time. Here overall fighting strength is the product of individual fighting value and the square of the number of combatants, a fact known as Lanchester's square law. One of the repercussions of this law is that a highly effective battle tactic is to have some means of dividing an opposing force into smaller groups, and then to take these on one at a time. Nelson is said to have won the Battle of Trafalgar by using this tactic.

Where Lanchester's linear law applies, individual fighting value is of great importance, but where his square law applies, number is of greater significance. This has been found to relate well to the different combat "strategies" of "slave-making" and "army ants".

Slave-making ants operate according to Lanchester's linear law. They attack colonies of other ants, either of the same or closely related species, and steal their brood. The captured brood (usually in the form of pupae) are brought back to the slave-maker's nest where they develop into workers, performing the same tasks that they would have done in their parental nest. Since slave-makers cannot possess numerical superiority, their ability to win depends on engaging defenders one at a time and having high individual fighting value, through being larger and better armed and armoured. They also produce "propaganda substances", which cause the defenders to scatter, or, as in *Harpagoxenus sublaevis*, to fight one another (cf. earlier discussion of the inquiline *L. kutteri*, which infiltrates the same host species, *L. acervorum* by this means). Either way, the effect is to prevent the defenders from mounting a concentrated attack on the invader.

Army ants operate according to Lanchester's square law. They are nomadic group predators that produce large colonies containing hundreds of thousands, or even millions of workers. The workers typically attack large arthropods or other social insect colonies. Relative to the individual army ant workers, these prey are large and dangerous, and indeed large numbers of army ants are killed daily in the course of combat. However, the prey are eventually overwhelmed by the concentrated attack which results from the huge numerical superiority of the army ants.

More generally, the mounting of a concentrated attack can be thought of as the result of a combination of mobility and the ability to focus effort at particular "pressure points" along contextual boundaries. An army which operates in this way therefore has the fluid-like properties which arise in all kinds of indeterminate systems, whether their individual components are identifiable as discrete entities or not. Also, the possession of indeterminate organisation can enable access to be forced into entities that are organized on a larger scale than individual attacking elements. To quote Sun Tzu, from 500 BC: "military tactics are like unto water; for water in its natural course runs away from high places and hastens downwards. So in war, the way to avoid what is strong is to strike what is weak. Water shapes its own course according to the ground over which it flows; the soldier works out his victory in relation to the foe he is facing".

4.2. Destroying Hosts

As has been mentioned, some parasites live in balance with their hosts, drawing resources from but not killing cells and tissues. For such parasites, localized death of the host may prevent access through such mechanisms as hypersensitivity, whilst generalized death allows other, less specialized organisms to establish.

For other, highly destructive parasites, which can be described as "necrotrophic", host death provides the means of access to resources. If they are to be evolutionarily successful these parasites must be able to proliferate from, survive and/or compete in dead host remains. They must also be able to exert sufficient "invasive force" to be able to overcome any passive or induced rejection mechanisms at their interface with the host. In the case of necrotrophic fungal parasites of plants, such invasive force has been called "inoculum potential". It depends on the fungus establishing a suitably well-resourced base from which it can maintain an invasion of plant tissues, which it destroys by producing toxins and plant cell-wall-degrading enzymes. If the base is insufficient, then the threshold potential required to overcome host rejection barriers will not be achieved and invasion will be curtailed.

For necrotrophic parasites to establish an effective base, they have to gather sufficient resources and concentration of biomass at the host interface. There are two basic ways in which this can be achieved. Firstly, the parasite may rely on events or situations outside its own sphere of influence. For example it may establish at locations where nutrient supplies are enhanced by exudation from the host and/or where host rejection is minimized—as in young, senescent or damaged tissues. Secondly, the parasite may use its own internal supplies, transferred from previous infections by means of migratory mycelium or survival structures such as thick-walled spores and sclerotia. In the case of migratory mycelium, this may be organized into cable-like aggregates or diffuse mycelial sheets or fans able to dilute out rejection in much the same way that these structures allow access to a neighbour's domain in purely fungal interactions. Once again, the balance between conservative, assimilative, explorative and redistributive processes is critical to success, as was described briefly for *Heterobasidion annosum* in Chapter 6.

4.3. Consuming Lesser Beings

As epitomized by such legends as Jonah and the Whale, any living entity which gains or is given access to the interior of a larger system is, unless very well insulated, in mortal danger of being digested, and so losing its inheritance along with its vitality.

This danger is evident in the fact that even mycorrhizal fungi (see above) can be digested. In those mycorrhizal fungi which "infect" orchids, the tables are completely turned, such that a would-be parasite-cum-importer of mineral nutrients and water becomes also the plant's means of assimilating carbon. The tiny seeds of all orchids actually need to be infected by mycorrhizal fungi if they are to germinate, and some orchids, such as the bird's nest orchid, *Neottia nidus-avis*, rely on the fungus to support their growth needs throughout their lives—as is clear from the fact that they do not produce chlorophyll. If sections through the orchid roots are examined microscopically, the outermost cortical cells can be seen to contain coiled fungal hyphae (known as "pelotons"), whereas the innermost cells contain digested fungal remains (Fig. 7.23). Interestingly, the mycorrhizal fungi of orchids include members of the genera *Rhizoctonia* and *Armillaria*, species of which are commonly necrotrophic parasites of other plants.

The example of orchidaceous mycorrhizas highlights the fact that whilst from the perspective of an invader, enclosure within another organism may provide shelter and access to resources, the livelihood of many organisms depends on their ability to ingest and digest others. By being able to break down the boundaries of ingested entities, the consumer gains energy at the same time as eliminating the potential threat of becoming infected or infested. This ability depends on either the death of the smaller entity prior to ingestion, or the containment of ingested entities before they can proliferate and/or inflict damage. Suitable partitioning devices range in scale from the food vacuoles of amoebae and gullets of *Paramecia* (see Chapter 2, Fig. 2.10) to the elaborate digestive tracts of vertebrates.

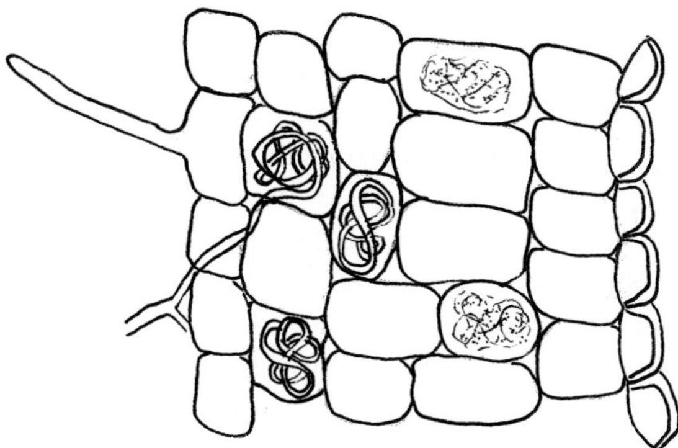

Figure 7.23. Structure of orchid mycorrhizas, as seen in a cross-section through part of a root from outer exodermis with root hairs (left) to the endodermis (right). Outermost cortical cells contain coiled pelotons, innermost cells contain digested hyphae.

Some means of recognizing or labelling ingested entities as other than self is also required if digestive processes are to find their appropriate target. At the cellular level, this can involve receptor-ligand interactions, as described in Chapter 2 in connection with phagocytosis. Proteins due for digestion in lysosomes are also generally labelled by a polypeptide known as "ubiquitin". The fact that ubiquitinization is involved with the turnover of dysfunctional self molecules raises the possibility that recycling processes within systems may also be responsible for elimination of nonself entities. However, what determines whether and when digestive processes are brought into action is not clear.

5. Rejection, Access and Evolutionary Instability

By now, the knife-edge balance between rejection, access and acceptance of other entities, and the varied consequences of these processes in the maintenance, integration and removal of contextual boundaries may be all too clear. Depending sensitively on circumstances, neighbours may compete or co-operate with one another and larger beings may consume, subsume or play host to smaller ones. Genetic information may be exchanged, interchanged, integrated or segregated. Populations may boom, bust or be maintained within bounds. Diversity, whether within individuals, species or communities may increase or decrease. Interdependencies may emerge or collapse. Complete integration, the combination of many into one, implies an inertness of being; absolute differentiation, the division of one into many, means total disintegration. Dynamic evolutionary processes depend on a continuing, complex interplay between rejection, access and acceptance.

The genetic exchange systems of basidiomycete fungi beautifully illustrate the principles underlying this complex interplay and its unstable consequences. As has been described, mating in these organisms involves fusion between sexually unspecialized hyphae and the invasion and acceptance of numerous nuclei within the boundaries of mycelial systems. There is therefore every potential for varied relationships to develop, depending on the degree of disparity between resident and incoming genomic populations, and for these relationships both to influence and be influenced by boundary properties. That this potential is actually realized is borne out by observations of the diverse ways in which mycelial systems of the same or closely related basidiomycete species can interact with one another in laboratory culture. To begin with, established heterokaryons containing populations of two kinds of nuclei with complementary mating-type alleles are generally inaccessible to ingress by additional kinds of nuclei. Correspondingly, heterokaryons typically express a rejection or somatic incompatibility response following fusion with other, genetically different mycelia, and multiply mated homokaryons become divided off into separate heterokaryotic sectors (see earlier this Chapter). From the point of view of genomic stability within heterokaryons, therefore, it would seem to be a case of "two's company, three's a crowd".

In keeping with their role as mating partners, basidiomycete homokaryons are accessible to invasion by nuclei having complementary mating alleles to their own. However, even when paired with a partner containing such nuclei, not all fusions result in access and indeed there may be significant expression of rejection prior to and even after access has been gained, as illustrated in Fig. 7.24. There is also evidence that access is more likely through hyphae that are developmentally more "juvenile", whereas rejection is more likely in longer-established hyphal systems.

Several studies have shown that the tendency of homokaryons to reject partners containing complementary mating alleles depends on the overall genetic disparity between resident and incoming genomes. In some cases, mates that are either related or very unrelated to the homokaryon are most liable to be rejected. This situation has parallels with many examples of mate choice in other organisms, where it has generally been viewed adaptively as a means of optimizing the benefits of outbreeding whilst reducing the costs of inbreeding or consorting with extreme non-kin. However, rejection in these situations can also be understood as the consequence of override systems that are insufficiently activated by the limited genetic difference between kin, but insufficient to prevent interference between genetically very disparate lines.

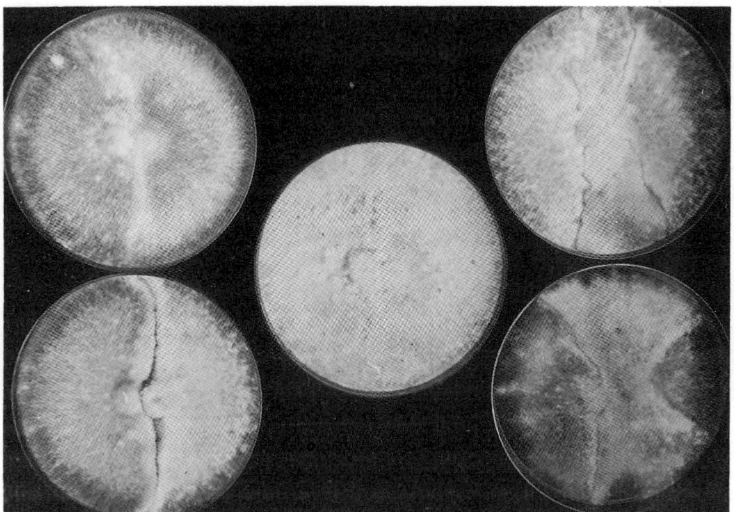

Figure 7.24. Love and war between siblings. Mycelia originating from different spores produced by the same "parent" fruit body of the basidiomycete fungus, *Stereum hirsutum*, have been paired next to one another in various combinations. Weak and strong zones of demarcation, due to rejection, have formed between the pairs shown on the left, which lack the complementary mating type genes that would allow them to mate. The pair in the centre possess complementary mating type genes and have fused and reciprocally exchanged nuclei to form a single "heterokaryotic" mycelium which occupies the whole dish. The pairs on the right also have complementary mating type genes, but their initial acceptance of one another has been superseded by rejection responses which confine the emerging heterokaryotic mycelium to central regions of the dish. (Courtesy of Dr A.M. Ainsworth).

Not all interactions between very disparate lines result in full rejection, however. Two examples of the extreme instabilities which can result from access between homokaryons from different continents (i.e. from "allopatric" populations) are shown in Fig. 7.25. These examples demonstrate the important point that as well as giving rise to reciprocal relationships, gaining or allowing access can lead to less mutually beneficial outcomes. The latter include extensive interference and degeneration, the suppression or physical replacement of one genomic population by another, and changing patterns of dominance from one partner to the other. In other words, sexual access may have as much to do with genomic territory as with the possible benefits of outcrossing. A parallel situation at the cellular level of organization has been found in colonial sea squirts. Here fusion between genetically different colonies has been found to allow "somatic cell parasitism"—the invasion of one colony by cells from another, and the development of gametes from the invader.

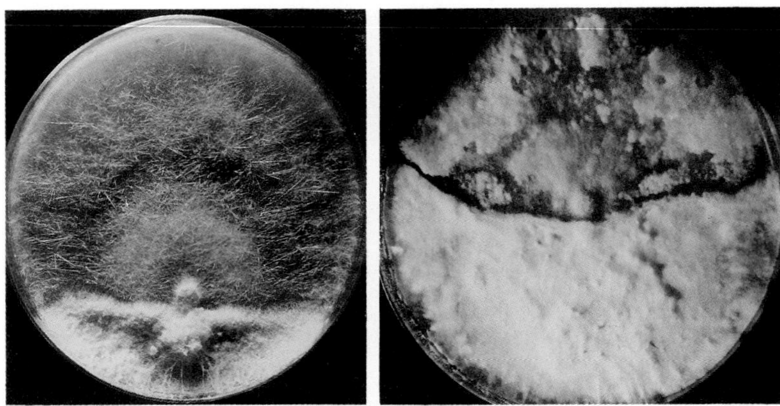

Figure 7.25. Unstable alliances between geographically isolated mycelia of *Stereum* species, when mated in Petri dishes in the laboratory. The dish on the left shows the degeneration of an Australian strain (inoculated uppermost) following reciprocal nuclear exchange with an English strain. Whereas the English strain remains vigorous, the mycelium of the Australian strain has collapsed and produced alternating zones of densely and sparsely branched crystalline filaments of the sesquiterpene chemical compound, (+)-torreyol. The dish on the right shows the complex outcome of reciprocal nuclear exchange between strains originating from the USA (inoculated lowermost) and what at the time of the experiment was the Soviet Union. Whereas the USA strain has remained vigorous, the Soviet strain has broken down into numerous, mutually incompatible subdomains expressing variable degrees of "western influence"! (From Ainsworth et al., 1990, 1992).

All these non-co-operative possibilities are prevented if rejection is expressed immediately between mating partners, and it is therefore of interest that the examples shown in Fig. 7.25 involve interactions between participants that would never normally encounter one another. Were such consequences of sexual access to occur within a population (i.e. "sympatrically"), very strong selection pressures would cause the elimination of the most susceptible by the most dominant members of the population, or the reinforcement of rejection barriers between sub-populations. Actual observations of the structure of basidiomycete populations within a geographical region indicate that such processes may indeed have occurred. Whereas some such populations consist of widespread, sexually non-outcrossing clones, inaccessible to genetic invasion, others consist of reproductively isolated "sibling species". The latter can be very alike to one another phenotypically, but exhibit a strong rejection response when mated.

If these "genomic relationships" within the protoplasmic labyrinths of basidiomycete mycelia sound uncannily close to human relationships within societies, then the aim of this book to emphasize the commonality of themes across organizational divides in living systems will be on its way towards being fulfilled. It now remains to consider what people, as perhaps the most self-consciously decision-making of all living beings, should or should not choose to do about it. To make love or to make war? That is the simple question with the complex answer.

CHAPTER 8

COMPASSION IN PLACE OF STRIFE: THE FUTURE OF HUMAN RELATIONSHIPS?

1. Pain

There is no escaping the fact that life contains at least some unpleasant moments for all human beings. Some pain is the inevitable consequence of our fragility as determinate, mortal individuals with indeterminate thoughts and aspirations.

However, a lot of human suffering is, quite literally, self-inflicted—the result of deep-seated attitudes towards ourselves and others that blight our ability to make the most of our lives together. Whereas advances in medicine and improvement of material conditions have been considerable for those people placed to take advantage of them, history records much less success in relieving the pain of social conflict. For the most part we have contrived to make life more difficult for one another than it need be, and continue to do so—by imposing unforgiving and unrealistic demands and expectations, if not by actual assault.

Whilst it may not be realistic or even beneficial to eliminate these attitudes, it may help to understand their origin and consequences and to appreciate when they are and are not apt. All too often they are attributed simplistically and dismissively to "human nature", which is an affront to ourselves as a species and an indictment of our understanding of living systems, of which we are both a part and an example. The punitive and retaliatory responses that we inflict on one another are a hypocritical, if expedient expression of the very same attitudes that these responses condemn, and can only re-inforce the vicious circle of abuse.

A biological perspective may help here by revealing the enormous variety of ways in which both the social and antisocial sides of human being can, when expressed in an appropriate context, contribute to a healthily versatile system. People are not unique in their social and antisocial drives; as I have strived to demonstrate, association-dissociation interplay is the universal consequence of living in dynamic boundaries. On the other hand, we may be uniquely self-conscious—prone to relate our personal welfare to the consequences of our own actions. We therefore tend to identify patterns of behaviour with the pursuit of goals rather than seeing them as the automatic outcome of organizational processes. At the same time, we may both be better placed and have more need to stand aside from the mêlée of immediate survival interests and thereby understand the origin of self-group tension and alleviate its more painful consequences.

When it comes to drawing lessons from biology, however, many people have eagerly grasped the principle that competition leads to individual and social betterment, and view co-operation as soppy and evolutionarily untenable altruism. Correspondingly, human compassion—fellow feeling—has become regarded as a noble aberration from the natural scheme of things—something which occurs, so to speak, in spite of ourselves and for which there is little rational justification.

Why have such ideas been accepted so readily? Is it because life is really like that, or is it because we all tend to see only those things that reinforce our prejudices and so use evidence selectively? Many people do after all harbour the suspicion, if not the certainty, that compassionate "softness" will result in a general "lowering of standards", or, worse, unleash all the "least desirable" elements of "human nature" and cause society to degenerate. Underlying such thinking is the belief, originating in a competitive, self-centred outlook, that no-one can really be trusted. This belief also leads on to the view that were it not for the imposition of all kinds of controls, civilization would collapse into a chaotic form of anarchy, an orgy of taking and no giving, with everyone out for themselves.

A competitive, mistrustful view of society goes hand in glove with a competitive, building-block, view of life. However, throughout this book, I have questioned the idea that patterns of life are based solely on competition between discrete survival units, pointing instead to the open-endedness and interconnectedness that comes from living in dynamic boundaries. I have also suggested that competition, when acting alone, will severely impede evolutionary innovation.

So, what if life is not after all solely about competitive struggle, and there really are powerful integrational forces that provide nurture in nature? Might this have a bearing on how we view the structure and dynamics of human societies? This is the issue that I want to explore in this final chapter, by trying to establish the extent to which the general biological principles I have been developing might be applied to ourselves.

I want to begin by examining the consequences of our mutual mistrust and the extent to which this mistrust has a real biological basis. I then want to show how the compassionate way out of the rut created by mistrust *emerges*, not as a weak, but as a strong option, founded on mutual regard and the courageous abandonment of old conceptual boundaries—old enmities.

I am only too well aware of the risks I am taking by discussing these issues. Some friends and reviewers (mostly scientists) have advised me firmly not to do it—that I should stick to what I know about biological systems and leave thinking about human beings to the "experts"! They have had several reasons for giving this advice.

First, there is the risk, already mentioned, of using biological evidence selectively in order to support a particular social or political viewpoint. All I can say in response is that this is not what I have set out to do. Rather, I have tried primarily to think about pattern-generating processes in all kinds of living systems and *then* to consider how well these processes might correspond with those that shape human societies. It is true that as a result of these considerations I have ended up with some strong opinions. However, I did not start with these opinions fully fledged; rather they have changed and strengthened as I have gone along—much as with many indeterminate processes—and they may yet change further.

Second, there is the risk of appearing to be inadequately "qualified" to comment on human affairs. I have been educated and trained as a biologist; I have not been taught sociology, economics, psychology or anthropology—and I have very limited knowledge of the literature and current thinking in these fields. However, I think that fearing to display personal insufficiencies—fearing to make a "fool" of oneself—is one of the greatest barriers to communication and mutual endeavour. All I am trying to do is describe the perspective that I have developed both as a biologist and from my own experiences.

The third risk, related to the second, involves the danger of seeing similarities between what are in reality very different systems. Many people regard humankind as a unique and unparalleled form of life, and so think that comparison with other life forms is not only a waste of time, but may also be dangerously misleading, if not insulting. For them, biology and sociology are entirely separate concerns and should be kept that way. However, I can see nothing wrong in principle with making comparisons as long as this is done without prejudice. If people really are unique, then the comparisons should reveal this to be so, just as much as they may reveal features in common where similarities do exist. I also think that when (and if) common features do become apparent, these can help understanding and provide some useful metaphors. I *do not*, on the other hand mean to indulge in "anthropomorphism", the *unquestioning* interpretation of the way organisms behave in terms of human motives—a much loved device used to popularize biology. Such interpretations are indeed prejudicial.

The fourth risk is that I am by no means the first to attempt to relate sociological issues to biological principles, and the path taken by my predecessors is littered with the remains of those (apart from those favouring arguments based on competition) who have succumbed to one or more of the other perils of the quest. Why, then, should my ideas stand any better chance of prospering? Maybe they don't, but past "failure" is no reason to stop trying so long as there appears to be good enough cause to do so. For me the need to follow up the repercussions of indeterminacy in a dynamic social context is a good enough cause, especially because the competition-based interpretations of biological and social systems have become so deeply ingrained and cause so much pain.

Finally, there is the risk that some of what follows may sound trite or self-important as well as smacking of the kind of Utopian idealism that has been so widely denounced over the years for its lack of realism. However, I have tried to base my thinking on facing up to all the attributes that human beings possess by dint of the fact that they *are living organisms*, rather than a desire to erase or deny those attributes which don't fit into some unrealistically "perfect" social model. Also, I make no claim of authority either over the ideas that I express or over the knowledge base from which these ideas have been induced. At best, I can only hope to stimulate some new lines of inquiry and re-stimulate old ones. Above all, I want to reinforce a fundamental message of this book, that there is more to be gained from the application of biological knowledge than the ability to control and exploit other organisms. Biological knowledge can help us to become more sensitive to the diverse qualities of life that both we ourselves and other organisms display as we interact with and create dynamic contexts in which to explore for, gather, conserve and redistribute resources.

So, here goes! Arm yourself with as large a pinch of salt as you think is necessary.

2. Sources of Strife—Parity, Disparity and Threats To Self-Fulfilment

Competition within and between human societies translates into fear. Fear results from the perception of a threat. Perception of a threat depends on the recognition of self-boundaries. Recognition of self-boundaries leads to possessive responses that preserve or extend self-domain. Possessive responses lead to strife. Strife leads to fear.

In this way, a vicious circle of fear perpetuates itself. However, like others, this circle can be broken if just one element in its sequence is removed. It is therefore understandable that many philosophies, religious or otherwise, have emphasized the abandonment of self-interest and/or possessions as a central theme. Death, the ultimate abandonment, is consequently seen both as a source of and a release from fear. On the other hand, the possibility of leaving something behind and thereby achieving some form of immortality eases the fear of death during life, but denies the solace that knowledge of the ultimate release of possessions, that all things must pass, can bring. Most perturbing of all is the notion of some kind of eternal afterlife of absolute bliss or absolute pain, before which each self is held to account for his or her social and antisocial actions (and thoughts). If ever a notion was designed to reinforce the fear cycle, serving the cause of social manipulators and coercers who exploit uncertainty at the expense of individuals' spiritual well-being, this is, or more hopefully was the one.

In whatever way the fear cycle may be maintained and reinforced, what gets it started is the inevitability that the presence of others within the same arena will result in competition and the threat of self-deprivation. Such deprivation is the ultimate cause of evolutionary extinction in all biological systems, and in human societies can take a wide variety of interrelated forms.

2.1. Deprivation of Resources

The most basic form of deprivation that competition leads to is a reduction in the availability of life-sustaining resources. For any living system or entity to thrive and survive it must possess a retentive boundary across and within which resources are transferred and sequestered. Where resources are freely available in the external environment, then this transfer can take place without immediately depriving neighbours, but sooner or later the resultant proliferation will result in a situation where supplies dwindle. By this time, and in the absence of integration, only those entities with the best-protected, best-defended boundaries will survive, and only those with the most invasive boundaries will prosper. Where there is disparity, the "weak" shall be dispossessed.

Such is the basic lesson that has been learned from studies of biological competition, and human beings are no more and no less immune to its consequences than any other living things. The subdivision of human populations by territorial boundaries as the outcome of competition for resources can therefore be understood in much the same way as can that of other organisms. Originally, the disputed resources would primarily have consisted of supplies of food and water, as they do in other animals. However, as the cultural feedback processes associated with "civilization" gathered momentum, an increasing range of natural and ultimately artifical and intellectual products would have become objects of dispute.

When discussing territoriality in Chapter 7, I implied that animal population numbers could be regulated at carrying capacity if each individual maintained a territory just sufficient to fulfil its needs for resources. Indeed definitions of animal competition often include some clause to the effect that competition only occurs when resource supplies fall below a threshold where they become insufficient to sustain individual needs. Above this threshold, there is surfeit.

The immediate inference from such reasoning is that if populations are held at or below carrying capacity, so that each individual receives all that it needs, then a stable, neutral co-existence can be maintained. However, an individual can only be satiated if it has determinate boundaries, and even spatially determinate individuals may fare better if they can assimilate freely in times of plenty so as to withstand times of shortage. Also, social groupings of determinate organisms have indeterminate boundaries, both in space and time. Living within one's means, at first sight a sure way to achieve strife-free, live-and-let-live societies is not so easily defined as might seem. Many view the inability of people to form such societies as the consequence of "human greed". However, greed—as an expression of insatiability—is inevitable in all systems that in one form or another can be considered to be indeterminate.

The fact that all human beings envisage an indeterminate future for themselves and/or their kin makes the kind of stability that might be attained by equitable distribution of resources, "to each according to his or her needs", well nigh impossible. Instead, dominance hierarchies ("pecking orders"), invasiveness and internecine strife will be the order of the day.

On the other hand, integrational processes that open up channels through which resources can be exchanged ("traded"), do provide scope for supply according to demand, complementation and access to a much wider array of resources. By joining trade networks, individual people or groups of people can therefore free themselves from the limitations of their individual territories and draw from a larger pool.

Despite these benefits, history reveals that the opening up of free trade also makes it possible to compete over much longer range, just as when fusion between fungal mycelia brings disparate populations of nuclei and mitochondria within the same protoplasm. Disputes across local territorial boundaries may then be amplified into global competition within network boundaries, leading to piratical and parasitic takeovers or to degeneration.

The invention of an informational system (money) to code for resources (rather as genes code for proteins), initially to ease trade by not having to lug actual commodities from place to place, has facilitated the emergence of socially parasitic, manipulative practices. These practices, such as stockbroking and running national lottery schemes, can make a profit out of redistribution without contributing in any way to the initial gathering and distribution of resources. They have gained social acceptance, even plaudits, by various kinds of subterfuge and propaganda—including such devices as causing resentment by dubbing those most in need as worthless, or paying dues to charity.

On the other hand, true investment can, and sometimes does, play an important role in the development of a redistributive infrastructure that directs resources to where they can be used most effectively. However, as money has lost its value as a token of productive human effort or natural bounty and become instead an instrument of power, investors have themselves become the target of propaganda from those competing to tap into their resource supply. Consequently, the effects of redistribution have often been to enhance rather than reduce disparity. Successful ventures have become monolithic establishments, locked in to their existing practices regardless of their real value to human communities, a large proportion of whose members become marginalized and labelled as failures.

2.2. Deprivation of Liberty

For any being with an uncertain future, the need to be able to direct its energies wherever the most favourable prospects lie is of paramount self-interest. The presence of others reduces this freedom, whether by getting in the way or by actual confinement, so that the threatened party feels "trapped". At the same time, a sure way of guaranteeing freedom is to restrict the freedom of others by producing or maintaining barriers and/or by uninhibited behaviour. There are all too many familiar examples of such restrictive and aggressive practices amongst people, paralleling those discussed in Chapter 7 for all kinds of living assemblages.

Integration with others—the ultimate "trap"—can be both profoundly restrictive and liberating. It can be liberating because of the emergence of properties in partnership that are denied to the individual. For example, not the least of the enormous changes in human circumstances that have occurred during the past century has been the opening up of prospects for travel and communication across and between continents, and even beyond the bounds of planet earth. On the other hand, there is a danger, remarked on in Chapter 4, that the opening up of infrastructure by joint enterprise can render the individual subservient to the growing demands of this infrastructure. Ultimately the reinforcement of infrastructural boundaries may so channel the activities of people as to make escape impossible; the rat race develops an interminable momentum and the human spirit seeks the nearest (or furthest) mountain top!

2.3. Limitation of Aspiration

Ambition, the ability to identify and desire to attain particular goals is something that can only truly arise in highly self-conscious beings. More generally, however, the need to maximize the potential to occupy niches is vital to the evolutionary success of all living systems, whether such goals are actually perceived by the systems themselves or not.

The realization of goals is crucially affected by the presence and responses of others, and in human societies, aspirations are often thwarted by competitors. The vast majority of us therefore have to settle for second best—the more so, the more that the development of infrastructure turns the whole world into a stage for those very few who attract the limelight. Concert pianists are rare, much to the grief of those whose budding talents and hard endeavours never get the reward of an hearing!

At the same time, the presence of others hugely increases the range of opportunities available to individuals, through social interactions.

2.4. Loss of Identity

An inevitable accompaniment to ambition is the desire for recognition. Recognition implies success, and therefore increased prospects of remembrance and/or genetic survival. To threaten someone's reputation is to threaten their indeterminacy. Integration with others, the becoming of a "mere" cog in the machine, can only be achieved with some loss of definition of self-boundaries. Such self-assimilation is greatly sought after as a means of evading exposure, but also greatly feared—as the route to self-annihilation. A feeling of belonging is one of the most reassuring sensations, but enforced anonymity is one of the severest forms of social deprivation.

Nowhere is the quest for recognition, and the accompanying fear of anonymity, more intense—or more pathetic—than amongst research scientists. Over the years, success in research has been defined by the impossible criterion of originality—being the "first" to make a discovery or conceive an idea. Being "second" is deemed worthless. Alternatively, success can be gained more securely by becoming an "authority", one whose real or apparent command of the knowledge base (or of the people possessing it) becomes widely known. Success, in whichever terms it is recognized, is rewarded, and therefore reinforced, with research funds.

The resultant scramble, rich as it is liable to be in piracy, parasitism, internecine strife, secrecy, deception, fraud and the pursuit of the obscure but safe detail neither makes an edifying spectacle nor the most effective use of resources. Getting out of the rutted battlefield is by no means easy, despite and perhaps because of the fact that many research scientists are extremely devoted to their subject. Although most scientists would claim only to wish to make a contribution to knowledge and/or understanding, their ability to do so and hence derive some sense of self-worth depends on recognition...

2.5. Incompatibility

Once the barrier of fear has been breached, and relationships have started to form, there is no guarantee that partnerships will survive and prosper. Indeed, so numerous are the ways in which mismatches can occur, that it is a wonder that any partnerships can endure. No sooner do incompatibilities cross a threshold (sometimes literally!) where they begin to outweigh the potential benefits of association, than they become amplified to the point of intolerable and irretrievable degeneration or outright dominance. This situation, which is generally the result of too much rigidity or too much flexibility, can be extremely costly to self-interest, sufficiently so to be a severe deterrent to forming anything other than the most superficial attachments.

2.6. Vulnerability

Sensitivity, the ability to respond to neighbours and surroundings depends on the possession of receptive and therefore relatively uninsulated boundaries. In human terms, the possession of such boundaries provides for an intensity of feeling which is the source of the greatest pleasure but also brings vulnerability in the form of dependency and ability to experience pain. Only the truly "thick-skinned" can bluster their way through life in "fast-forward" and trampling roughshod over others—in much the same way as a non-assimilative mycelial phase will smother an assimilative one (see Chapter 7). Having experienced pain—once bitten, twice shy—all sensitive beings tend to become self-protective, and hence thick-skinned themselves. How else can a person become inured to scenes of violence delivered to them daily by news media? How, once an entire society has become thick-skinned can its members be capable of mutual aid?

3. Self-perpetuating Consequences of Strife

Any system that continues to be structured solely on the basis of competition will remain strife-torn, subject to all the threats to self-fulfilment that I have described. In human societies, the existence of competition in itself has a whole set of consequences which serve only to perpetuate strife. Not the least of these consequences is the development of attitudes of mind which regard all competition as healthy, the sure way to promote ambition and advancement—notwithstanding its cost in a few lost souls. Needless to say, these attitudes are most readily espoused by those who actually succeed (or think that they succeed), so gaining executive influence over their fellows.

3.1. Success and Failure—Winning and Losing

In a competitive system, success breeds success. It also breeds failure. For every winner there has to be at least one loser (and sometimes many more). The traits that bring success may be passed on by genetic or other means of information transfer, but the recipients of these traits only end up competing with one another. Unless a monopoly develops or there is some kind of niche subdivision (see Chapter 1), winning lines inevitably become losing lines in the end.

From winners' perspectives, however, the traits that have got them where they are in human societies seem admirable and well worth promoting. In the midst of triumph, the last thing on winners' minds will be their own long-term humiliation or the cost that their victory exacts from the vanquished. Yet any dispassionate spectator who has watched winner take all in some football cup final or election would be thick-skinned indeed not to notice the desolation of the losers nor appreciate the fine twists of circumstance that can make the difference between defeat and victory. By the same token, anyone who has noticed that after months or years of preparation the difference between Olympic gold and silver may be as little as one hundredth of a second, is likely to have wondered at the disparity of treatment afforded to the recipients of the medals.

3.2. Refinement and Dissipation

The Olympic example epitomizes the costs and benefits of competition in human societies. On the one hand, competition provides a stimulus for striving to enhance performance in pursuit of a particular objective. If that objective is attained, the effort may seem to be worthwhile, at least for the victor, and from a dispassionate standpoint even losers play an important role in extracting the very best performance out of eventual winners. On the other hand, gains have to be balanced against costs and from an overall perspective the additional effort required to provide a minute advantage as competition intensifies, often borders on absurdity. Competition may indeed aid refinement when a field is relatively open, but as limits are approached it becomes increasingly dissipative, especially as the participants become prone to various kinds of "cheating" and incentive turns to disincentive. At this point, only the unpredictable intervention or emergence of a participant with a fresh approach, free from the entrenching effects of maintaining short-term competitiveness, can save the field from stagnation. Such participants are invariably resented at first, then copied, then give rise to their own competitive cycles and demise.

3.3. Notions of Superiority and Inferiority—Self-sufficiency and Insufficiency

A pervasive inference from the fact that competition generates winners and losers is that there is disparity in human societies between individuals who are better or worse than one another in various respects. Furthermore, the interpretation of the Darwinian message to the effect that success means having "better" genes has encouraged the notion that differences in peoples' abilities are largely inherent. In these terms, success and failure are inevitable correlates of individual worth. Those who achieve success therefore feel, or are made to feel, superior, and gain in ambition and responsibility. Those who fail, feel, or are made to feel, inferior and lose ambition and responsibility.

Notions of success and failure have in turn become entwined with concepts of self-sufficiency and insufficiency—whether an individual is a fully-fledged, independent and valued member of society or a dependent inadequate. As the demands of an increasingly complex life style have increased with the development of supposedly supportive technologies, so the expectations on individuals to remain self-sufficient by increasing their range of skills and activities has escalated. "Educational" establishments call for longer and longer periods of "training" before the individual is "fit" to be launched into the world of "skilled, responsible work". Inability to cope becomes an insurmountable problem as stress (=expectation) levels exceed the thresholds of individual tolerance. Yet, to admit to this inability is to admit "failure" and risk consignment to the growing pile of human debris—"no-hopers" who only vicariously can taste the fruits of fulfilled ambition.

3.4. Hero-worship and Inequality of Status

The rare few who make it to the top of highly competitive fields in human societies find themselves in an unenviably enviable position. They receive rewards colossally in excess of their immediate needs for resources, and influence well beyond their immediate neighbourhood. Powerful positive feedback loops heap success upon success as winners become heroes—targets for the jealousy of rivals and adulation of the unfulfilled.

The emergence of hero-figures has been a recurrent theme in science. Whether the individuals concerned would have wished it or not, they first become foci for the initiation of movements and then become stumbling blocks as their disciples vigorously defend their authority. Eventually, the inevitable weaknesses in hero figures begin to show and to be amplified in the glare of exposure until their whole facade begins to crumble. At this stage, the danger of annihilation, the destruction of all that might have been sound as well as unsound, becomes great. If "fortunate", the hero (or rather the hero's reputation) may survive this stage, eventually joining the ranks of flawed geniuses upon whom history bestows the honour of having contributed to the advancement of knowledge. If not, only oblivion or perpetual ridicule lie in store.

All this seems rather extreme when it is realized that with some notable exceptions heroes generally emerge as the product of utterly unpredictable feedbacks rather than being bred and groomed for the role. For although when the career of a hero figure is reviewed it may seem that all was somehow predestined, a conspiracy of events and the innate qualities of the central character, that perspective is the classic outcome of hindsight. Trajectories that are unpredictable with foresight can always be rationalized with hindsight. Also, if the trajectories are analysed not in terms of their actual outcome, but rather in terms of what might have been, then the enormous influence of serendipity—emerging in the right context—becomes apparent.

With the widespread acceptance that success is the reward for admirable intrinsic qualities, both in evolutionary systems in general and human societies in particular, the importance of serendipity—and its counterpart, misadventure—has been much neglected. The reasons for this neglect may lie partly in the building-block treatment of populations as equilibrium systems that are large enough for chance effects to be averaged out, and partly in the fact that being so unpredicatble, serendipity is very hard to quantify. However, the potentially overwhelming effect of serendipity in an indeterminate, nonlinear system may be evident from a re-examination of Fig. 3.2 (Chapter 3), showing the mycelium of the magpie fungus growing in a matrix. All the non-assimilative hyphae distributed across a low nutrient field have equal genes and equal opportunity. Those that happen to locate the connections onto the high nutrient domains are the ones which become centres of attraction for followers, clearly demonstrating how leaders can be made by circumstances and not by themselves.

3.5. Centralized Power

The idea that there can be a selectable "best" for leadership of human societies, combined with the development of infrastructural networks within which disparities are amplified by competition rather than equilibrated by co-operation, leads inexorably to hierarchies and centres of power. At these centres are to be found managers, chairpeople, field marshals and, in the case of nation states, Heads of Government, through whose Offices all decision-making processes must ultimately be fed.

Throughout much of human history, such centralized power has been held by many to be the most efficient way of administering order and coherence in societies, without which all would collapse into chaos. Leaders themselves have done little to disabuse others of this idea, even though they themselves may suffer the appalling stresses of input overload and enmity. Even if they were to try to do so, they would probably not be believed; the logic of central administration seems so irrefutable. For example, doesn't the functioning of our own bodies demonstrate the efficacy of central control all too clearly? At heart there is a pulsating muscle which gathers and drives fluid around the blood circulation. Regulating the operation of the endocrine (hormonal) system is the pituitary gland. Seemingly in control of cell form and functions, and from there all other bodily properties, is the nucleus with its executive "committee" of genes. Perception and response to environmental inputs are effected through a "central" nervous system, headed (literally) by the brain.

However, administering a determinate system, such as a human body, is not the same as administrating an indeterminate one, such as a human society. Also, first impressions can be deceptive. For example, recent research suggests that there is no organizational centre of consciousness in the brain; instead, streams of consciousness flow through a parallel-distributing network in the cerebral cortex.

The fundamental tenet that "executive heads" provide the only conceivable means of co-ordinating societies may therefore need some re-examination. In fact, when the issues are thought through, some serious problems become apparent. Firstly, there is no good reason to believe that leaders are superhuman, or indeed subhuman. Like anyone else they will possess strengths and weaknesses as well as the particular talents which for one reason or another make them centres of attraction at the time of their selection. To expect them to be omniscient arbitrators of all issues is therefore ludicrous and bound to cause trouble. Secondly, if everything has to be referred back to central channels, the resulting log jam will greatly impede any decision-making process. Thirdly, peripheral regions at the dynamic boundary of a system—those whose responses to circumstances are most critical to future developments, will be just those that are most out of touch with the centre and *vice versa*.

3.6. Intolerance of Non-conformity

For centres to assert their authority, the systems that they govern must operate like clockwork—precisely, predictably and without error. Any actual decision-making has to be reduced to a minimum, and where such decisions are enforced, they will be those which best maintain the *status quo*. Non-conformity, whether in the form of rebellious individuals, ideas or errors threaten to disrupt the *status quo* and cannot therefore be tolerated. Laws, codes of conduct, militia, policing and various modes of incarceration and deprivation of resource supplies are therefore essential instruments of central power. When things do not go to plan, the errant individuals must be sought out, blamed, and made an example of.

3.7. Homogenization

If competition is to lead, as it does, to monopolies of power and intolerance of non-conformity, then the inevitable consequence is the homogenization of societies into subservient, unquestioning flocks of sheep (with apologies to goats) that toe the established line. Freedom of expression within such societies is restricted to the choice of predetermined careers; new ventures are impossible. At the boundaries of these societies, encounters with others with different viewpoints pose a considerable threat (and promise) of instability...

3.8. Incentives and Career Structures

In a clockwork society, there are jobs to be done if the existing power structure is to be maintained. Since individuals are determinate, a continuing supply of recruits must be trained in the ways of fulfilling the many diverse operations required to keep the system ticking. The competitive hierarchy that has made the clockwork then becomes, in itself, the recruiting system, luring callow workers with the possibility of success, frightening them with the possibility of failure. For this system to be effective the disparity between relative success and relative failure must be great. Without the dream of rare success and vast reward—and nightmare of poverty—there can be no incentive in a non-caring, non-sharing, unforgiving society to work for work's sake, dissipating resources on the grand scale and with no time to appreciate anything on the wayside.

3.9. Continual Assessment

For competition, the blind clockmaker, to keep a system in order, there must be plentiful opportunity for selection of those best fitted for particular positions within the hierarchy and elimination of error. Recruits must be put through trial after trial; those in post must be continually assessed to make sure they are doing their jobs—for in a competitive regime no-one is to be trusted, least of all those doing what they do for love! The fact that more time and energy may then be spent in assessment than in doing or not doing anything is just one of those things. People do not test a car to destruction (they only test a replica) before putting it on the road, but they certainly do it to themselves.

3.10. Self-satisfaction, Stress, Anxiety, Depression and Guilt

The overall emotional costs of so much competition, so much hurdling, are terrifying. For those who succeed, or think they succeed, there may be some self-satisfaction, accompanied by insensitivity and intolerance towards others. On the other hand, the pleasure of having met a challenge may be tempered by the guilt of having deprived others in the process. Such guilt may be amplified in those who know their own weaknesses and therefore wonder about the "fairness" of their triumph. For those who fail, or think they fail, or think that they should have failed, the bottomless pit of depression beckons. For those who are blamed, or blame themselves, both guilt and depression tug for authority, notwithstanding the fact that fault in individuals and fault in societies are difficult to separate. The only protection from guilt and depression is callousness, the acquisition of a thick skin and contempt for others. For "successes" and "failures" alike, stress and anxiety are amplified by society's unending demands and the feared exposure of inadequacy.

3.11. Immobility

The clock ticks. Time passes. Nothing changes.

4. Foundations of Compassion—Mutual Supportiveness

The picture just painted of a purely competition-based human society is bleak. Fortunately, it does not fully represent the way human social life has developed, nor, hopefully, how it will develop in the future. Co-operation, the antithesis of competition has always been very important in human relationships. However, just as with competition, it is possible to have too much of it, so that it stifles, just as much as competition dissipates, energy. The question that therefore arises is how best can the divisiveness of competition be counteracted to produce societies in which avoidable suffering is prevented whilst the abilities to explore, innovate, keep in touch and, above all, enjoy life are sustained?

4.1. Self-extension—You are My Concern, They are Our Concern

The first step in the emergence of compassion comes with the conscious realization that the boundaries of self and of neighbours are all vulnerable and not as separate from one another as might at first seem. True, nobody can directly sense the pain or joy felt by another, but all can extrapolate from self-awareness to the (almost) certain knowledge that such common pain and joy exist. Similarly, awareness of self-needs and the fact that others can help to meet them brings a corresponding awareness of the needs of and ability to help others. Common pain, common joy and common interest are common knowledge, the bases for common passion, sharing and caring. Vulnerable self boundaries become merged, but not submerged, with those of neighbours; there is fellow feeling, but not subservience.

4.2. Identifying a Common Purpose

Not surprisingly, compassion is most evident when some external threat, beyond individual control, looms large, as in wartime or in the presence of insuperable natural forces. At such times, when all are in the same boat and none can survive alone, self boundaries lose their definition as individual and collective responsibilities coalesce and self-concerns cease to serve any useful purpose. A communal thrill is experienced as each individual becomes absorbed into the whole, accompanied by an almost palpable release of energy as the effort to maintain self-boundaries is relaxed. The popularity of war and disaster movies, social drinking or drug-taking and all kinds of gatherings at "events" owe much to this release of energy, as do rulers who use it as a means of deflecting their subjects from self-concerns. Fusion energy is not confined to the world of molecules and atoms! The formation and maintenance of boundaries involves energy input and so the integration of boundaries allows energy release.

The tricky bit is to get this compassionate mechanism to work during times when the external environment seems bountiful and self-boundaries and competitive drives are prone to reassert themselves. Some generally applicable common purpose needs to be identified. I can see no better way of achieving this than using our self-consciousness to view the prevention of avoidable human suffering and enhancement of quality of life as a common purpose, and to identify excessive competition as a common enemy.

To re-iterate, this does not mean that all competition is fearful; in moderation, competition can aid differentiation, refinement and eventual co-operation through division of labour. There is a threshold before which competition is healthy and beyond which it is not. The task is therefore to identify the circumstances and attitudes which cause this threshold to be exceeded, and to recognize the danger signs that it is being exceeded.

In a few glib words, the predominant causes of excessive competition are overpopulation, the equation of self boundaries with individual boundaries and too much infrastructure. The predominant symptoms, in no particular order, are:- warfare, fear of failure, crime, punishment, callousness, emergence of centres of power, blame, conceit, bigotry, secrecy, authoritarianism, oppression, subservience, environmental destruction, fault-intolerance, hero-worship, examination systems, mistrust, social parasitism, ridicule, demotivation....

So, what preventive, curative and palliative treatments are available for the disease of social existence?

4.3. Maintenance of Communication Channels

All co-operation depends on communication between participants. In human terms, the ability to relate individual concerns, feelings and thoughts—by talking, writing, facial expression, body language or whatever—is fundamental to the formation of mutually supportive relationships. At the same time, the absence or breakdown of communication and the resultant maintenance or reinforcement of boundaries is the first sure step towards the development of me-and-you, them-and-us mentalities. Opening communication channels, and ensuring that they remain open, or at least openable, is therefore fundamental to a thriving co-operative society.

4.4. Wariness of Stifling Infrastructures

At the same time as maintaining the scope for communication it is vital not to dissolve self-boundaries altogether. If there is no partitioning, there is no degree of freedom within the system and individuals lose their aspirations along with their self-identity. The scope for complementation then diminishes, huge power drains can develop, energy cycles round the system, but nothing gets done. In short, the system embeds itself and its inhabitants in a rut. Standing up for individual rights, not allowing oneself to be taken in, in other words loving one's self as well as one's neighbours, is as important a part of living in a compassionate society as the willingness to open barriers and help others.

4.5. Respect and Preservation of Differences—Avoiding Homogenization and Allowing Complementation

To conform may be the easy way of avoiding conflict, but it is not the best way to contribute to the common wealth of spirit and resources. In fact it is liable to lead unerringly to disaster as monopolizing competitive processes are allowed free rein. Differences need to be respected and preserved if there is to be the real complementation that enables emergence to new vistas of opportunity.

4.6. Aspiring to Roles—to Competence and a Sense of Belonging Rather than Competitive Success

Without competition, disparity, assessment and selection it is commonly argued that there would be "no incentive" for individuals to "improve themselves" and thereby enhance their contribution to society. Certainly virtually all "educational" establishments abide by this dogma and continue to crack the whip of examination systems.

Not only does this dogma cause untold suffering, but it is actually counterproductive in a wide variety of ways. It reinforces the attitude that all competition is good. It is liable to cause as much, if not much more disincentive amongst those who fail (for whatever enormous variety of reasons) or whose insights cause them not to wish to succeed, than it provides incentive. It necessitates some kind of terms of reference for assessors (examiners, interview panels, promotion boards, shareholders etc) that enables them to differentiate "better" from "worse". The problem here is that there are some qualities, notably those associated with creativity and co-operativeness, which are extremely difficult to quantify compared with others such as industry, remembrance of facts, performance of routines and freedom from errors. It is therefore the latter, more quantifiable characteristics—

which are best enhanced by "training" (see Chapter 3)—rather than the former characteristics, the product of "education", which provide the predominant basis for "qualifications". Re-iteration becomes the name of the game. Also, where suitably quantifiable criteria are used for assessment, then the competitive need to do better in order to be successful sets a mindless treadmill in motion. The treadmill operates on the basis that "more is better"; participants endeavour to outdo one another in sheer weight of work or multiplicity of technical skills, regardless of relevance or utility. The soul-destroying, relationship-destroying, work-for-work's-sake syndrome becomes entrenched.

None of this would be necessary if it were not for the cynical notion that left to themselves most people would rather do nothing. Personally, I haven't met any such people, except those who are ill or exhausted, though I have met many who would rather conserve energy than do anything boring *and* (*not* or) pointless. Interest and relevance are the key to motivation, and it is these that education should endeavour to foster, not the dispiriting message that "you can only succeed if you outwit your peers and don't err". As far as relevance goes, the message that "by developing this talent or by doing this service you can make life more enjoyable for yourself and for your friends and feel you belong" is likely to select far more positively than "if you don't do, and don't do it better than anyone else, you won't get". People can be encouraged towards wanting to find a role, performing it to best effect, and being rewarded by the benefits to themselves and others that it brings. The desire to do nothing is the consequence of a sense of individual powerlessness in the face of competition and coercion, not the reason for instigating these latter.

4.7. Self-appraisal

How, then, can individuals find suitable roles to aspire to? In order to do this they need to develop interests and to have some unbiased way of appraising their own strengths and weaknesses (however these might arise) relative to others. This is where educational systems and an element of competition can come into their own, as the means of providing guidance and feedback, but not judgement. The individual can sample various options, so gaining the general background necessary to interface with others, before selecting particular pathways and developing the skills necessary to follow them. There are many signs that this route towards self-appraisal and selection, as opposed to the imposition of qualification barriers, is one that many educationalists would like to follow. Whether those empowered by competitive success to be decision-makers would ever allow it to happen is another matter.

4.8. Equality of Status

One thing is certain: so long as there are huge disparities in reward and status associated with different roles in society, no method of self-appraisal as the means of making selections could work, because it would be impossible to avoid bias. The reduction of these disparities, for all that it might be deemed to drown competitive incentive, is essential for co-operative societies to develop and function. This does not mean to say that there should be no disparity, since some roles require more effort and so deserve more reward, but only that the excessive inequalities of the present day should become a feature of the competitive past.

4.9. Distributed Power

An important source of the large disparities in human societies is the development of centres of power that result from the operation of competition within infrastructural networks. As has been mentioned elsewhere, there are important reasons for regarding such centres as detrimental to the functioning of societies as versatile, responsive systems.

Mechanisms that decentralize power stop anything or anyone getting "too big for their boots" and at the same time allow collective decision-making to be done at local level, where it counts, in a distributed power structure. The main mechanism which allows this kind of structure to develop is the give and take of neighbourliness, attuning one's behaviour to correspond with and, where appropriate, counteract, those in the immediate vicinity. The abilities for responses between neighbours to be amplified into coherent phenomena affecting an entire system is demonstrated by the fluid dynamic properties of flocks of birds and shoals of fish (see Chapter 3), not to mention the networking properties of fungal mycelia (Fig. 3.2). At the same time, the limitation of communication within local channels reduces the danger of power drains developing and maintains the scope for all to play a part. There is no need for governors and the accoutrements of central power. Anarchy, in its profoundest sense as a coherent co-existence by self-government becomes at least a theoretical possibility if (and its a very big "if") competitive influences can be kept within bounds.

4.10. Respect and Tolerance of Error and Insufficiency

In a distributed power structure, errors and inadequacies are much more easily circumvented and much less catastrophic than in a centralized system. Distributed power therefore allows individual insufficiency to play its rightful part in the development of complementary relationships. It also allows the freedom to err which is essential to exploration and adjustment to changing circumstances. Fear of failure and the burden of colossal responsibility are greatly reduced, whilst the inspiration of being able to participate, to exercise one's strengths and override weaknesses, to support and be supported, is retained. Furthermore, when and if the need arises to redeploy from redundant to emerging activities, then the support of the community can aid in the abandonment of old boundaries and readjustment.

4.11. Combining Private Enterprise with Socialism—Balancing Conservation, Exploration, Assimilation and Redistribution

Please look once again at Figures 3.2 and 6.1. By differentiating and integrating, reconfiguring and recycling, a distributed power structure in which boundaries are opened or sealed off according to circumstances allows a versatile indeterminate system to develop. This system gathers and stores, explores and redistributes within its dynamic boundaries as its components assert their autonomy and admit their interdependency in a continuing creative interplay.

4.12. Fluidity

Patterns emerge and change. Current flows. There are degrees of freedom.

REFERENCES

The literature listed below is divided into two sections.

The first section, "Suggestions for Further Reading" is a small selection of recently published books which contain ideas and/or information that can usefully be compared with what I have suggested and/or stated in "Degrees of Freedom". These books can also, if desired, be used as a source of additional literature on evolutionary biology, including the influential writings of Richard Dawkins, Stephen Gould and Lynn Margulis, amongst others. Neither the presence nor the absence of a title in the list should be taken to imply any judgement on my part about its contents.

The second section—"Literature Cited in Figure Captions"—is just that, and can be used to track down sources of specific factual details and concepts.

Suggestions for Further Reading

J.H. Andrews, *Comparative Ecology of Microorganisms and Macroorganisms* (Springer-Verlag, New York, 1991).

L.W. Buss, *The Evolution of Individuality* (Princeton University Press, Princeton, 1987)

J. Cohen and I. Stewart, *The Collapse of Chaos* (Penguin, London, 1994).

P. Coveney and R. Highfield, *The Arrow of Time* (Flamingo, London, 1991).

E. de Bono, *I am Right—You are Wrong: from this to the new renaissance: from rock logic to water logic* (Viking, London, 1990).

D.C. Dennett, *Consciousness Explained* (Little, Brown, Boston, 1991).

B. Goodwin, *How the Leopard Changed its Spots* (Weidenfeld & Nicolson, London, 1994).

G. Johnson, *Fire in the Mind* (Viking, London, 1996).

S.A. Kauffman, *The Origins of Order* (Oxford University Press, Oxford, 1993).

H. Norberg-Hodge, *Ancient Futures* (Sierra Books, San Francisco, 1992).

C.D. Rollo, *Phenotypes* (Chapman & Hall, London, 1995)

I.K. Ross, *Aging of Cells, Humans & Societies* (Wm. C. Brown, Dubuque, 1995).

S.C. Stearns, *The Evolution of Life Histories* (Oxford University Press, Oxford).

M.M. Waldrop, *Complexity* (Simon & Schuster, New York, 1992).

Literature Cited in Figure Captions

A.M. Ainsworth, in *Evolutionary Biology of the Fungi*, ed. A.D.M. Rayner, C.M. Brasier and D. Moore (Cambridge University Press, Cambridge, 1987), 285-299.

A.M. Ainsworth *et al*, *Mycol. Res.* **94** (1990) 799.

A.M. Ainsworth and A.D.M. Rayner, *Mycol. Res.* **95** (1991), 1414-1422.

A.M. Ainsworth *et al*, *J. Gen. Microbiol.* **138** (1992), 1147-1157.

R.C. Aylmore and N.K. Todd, in *The Ecology and Physiology of the Fungal Mycelium*, ed. D.H. Jennings and A.D.M. Rayner (Cambridge University Press, Cambridge, 1984), 103-125.

R.C. Aylmore and N.K. Todd, *J. Gen. Microbiol.* **132** (1986), 581-591.

C.M. Brasier, *Adv. Plant. Pathol.* **5** (1986), 55-118.

J.L. Burton and N.R. Franks, *Ecol. Entomol.* **10** (1985), 131-141.

M.J. Carlile and J. Dee, *Nature* **215** (1967), 832-834.

M.J. Carlile, in *Methods in Microbiology vol. 4*, ed. C. Booth (Academic Press, London, 1971), 237-265.

M.J. Carlile, *J. Gen. Microbiol.* **71** (1972), 581-590.

M.J. Carlile and S.C. Watkinson, *The Fungi* (Academic Press, London, 1994).

D. Coates *et al*, *Trans. Br. Mycol. Soc.* **76** (1981), 41-51.

D. Coates and A.D.M. Rayner, *New Phytol.* **101** (1985), 183-198.

R.C. Cooke and A.D.M. Rayner, *Ecology of Saprotrophic Fungi* (Longman, London, 1984).

C.G. Dowson *et al*, *J. Gen. Microbiol.* **132** (1986) 203-211.

C.G. Dowson *et al*, *FEMS Microbiol. Ecol.* **53** (1988a), 291-298.

C.G. Dowson *et al*, *New Phytol.* **109** (1988b), 423-432.

C.G. Dowson *et al*, *New Phytol.* **111** (1989), 699-705.

N.R. Franks, in *The Collins Encyclopedia of Animal Ecology*, ed. P.D. Moore (Equinox, Oxford, 1986), p.46.

N.R. Franks and P.J. Norris, *Experientia Supplementum* **54** (1987), 253-270.

G.S. Griffith *et al*, *Nova Hedw.* **59** (1994a), 47-75.

G.S. Griffith *et al*, *Nova Hedw.* **59** (1994b), 331-344.

J. Heslop-Harrison, *Cellular Recognition Systems in Plants* (Arnold, London, 1978).

E.B. Lane and M.J. Carlile, J. Cell Sci. **35** (1979), 339-354.

A.D.M. Rayner and N.K. Todd, *J. Gen. Microbiol.* **103** (1977), 85-90.

A.D.M. Rayner and L. Boddy, *Fungal Decomposition of Wood* (Wiley, Chichester, 1988).

A.D.M. Rayner *et al*, *Can. J. Bot.* **73** (1995), S1241-S1258.

A.D.M. Rayner, in *A Century of Mycology*, ed. B.C. Sutton (Cambridge University Press, Cambridge, 1996a), 193-232.

A.D.M. Rayner, in *Fungi and Environmental Change*, ed. J.C. Frankland, N. Magan and G.M. Gadd (Cambridge University Press, Cambridge, 1996b), 317-341.

A.D.M. Rayner, in *Aerial Plant Surface Microbiology*, ed. C.E. Morris, P. Nicot and C. Nguyen-The (Plenum Press, New York, 1996c), 139-154.

J.M.W. Slack, *From Egg to Embryo*, (Cambridge University Press, 1991)

R.J. Stipes and R.J. Campana, eds, *Compendium of Elm Diseases* (American Phytopathological Society, St Paul, Minnesota, 1981), p. 12.

J. Webster, *Introduction to Fungi* (Cambridge University Press, Cambridge, 1980).

E.O. Wilson, *Success and Dominance in Ecosystems: The Case of The Social Insects* (Ecology Institute, Oldendorf/Luhe, Germany, 1990).

A.M. Woods and J.L. Gay, *Physiol. Pl. Pathol.* **23** (1983), 73-88.

INDEX

Numbers in bold type refer to pages on which relevant illustrations appear.